The Strong, Gravitational

& Electric forces

by
Vladimir B. Ginzburg

Edited by Ellen Orner
Cover by Eugene Ginzburg

First Edition

Helicola Press
Division of IRMC, Inc.
612 Driftwood Drive,
Pittsburgh, Pennsylvania 15238

Library of Congress Control Number: 2002111421
ISBN: 0967143217

Current printing (last digit):
10 9 8 7 6 5 4 3 2

*To my wife Tatyana, my daughter Ellen,
my son Eugene, and granddaughter Alexandra.*

That the man calls matter, or substance, has no existence whatsoever. So-called matter is but waves of the motion of light, electrically divided into opposed pairs, then electrically conditioned and patterned into what we call various substances of matter. Briefly put, matter is the motion of light, and motion is not substance. It only appears to be. Take motion away and there would not be even appearance of substance.

Walter Russell.

. . . Yet we can not rest satisfied until the deeper unity between the gravitational and electric properties of the world is apparent. . .

Sir Arthur Eddington.

PREFACE TO THE SECOND PRINTING

After publishing the first printing of this book in August 2002, I received a number of valuable and constructive comments from professional physicists in response to my brief description of the model related to the unification of the strong and electric forces. I would like to thank the following gentlemen:

Dr. Rudolph Hwa, Professor, University of Oregon

Dr. Edward F. Redish, Professor of Physics, University of Maryland

Dr. David F. Measday, Professor of Physics and Astronomy, University of British Columbia, Canada

Dr. Eric Carlson, Wake Forest University

Dr. Warren Siegel, Professor, Stony Brook University

Dr. Charles E. Hyde-Wright, Professor of Physics, Old Dominion University

Dr. John G. Fetkovich, Professor Emeritus, Carnegie-Mellon University

Dr. Yan Corlett, York University, Canada

Dr. Igor Sokolov, Assistant Professor, Clarkson University

Dr. Thomas Love, California State University at Dominguez Hills

Dr. Harold Fox, Editor, The Journal of New Energy.

Their comments encouraged me to expand further the spiral field theory and to demonstrate a correlation between the predictions of this theory and experimental data. As the reader will find in this book, this theory predicts accurately, among other things, the peculiar magnetic moments of the nucleons that have not been satisfactorily explained by current theories of nuclear magnetism.

Vladimir B. Ginzburg

Contents

PART 1

*Spiral Nature
of the Universe*

A

ABOUT THE BOOK

Ten Major Points

We live in a world that is spiral all the way from inside out. All the way, from micro-entities such as elementary particles, nucleons, atoms, DNAs, neutrinos, and electromagnetic waves to macro-entities that include tornadoes, hurricanes, planetary systems and galaxies. Although this proposition was repeatedly made by many scholars since the beginning of sciences, only nowadays we began to recognize its profound significance. This is thanks to the works produced by a number of scientists during the 20th century and at the outset of the new millennium. Among these works is the spiral field theory described in this book.

The great importance of geometry in nature was recognized a long time ago by the ancient Greek scientists, and since then geometry became present in all the theories of physics. Remarkably, out of all possible geometrical shapes available in nature, it was the spiral shape that was to play an exclusive role in both the construction of our universe and sustaining its existence for billions of years. The spiral field theory fully subscribes to this assessment of the unique role of the spiral shape in nature. Moreover, it shows that by recognizing the spiral character of Nature, one can get even closer to discovering some of Her sacred secrets. Here are the key discoveries of the spiral field theory.

1. Discovery of the unique multiple-level spiral shape called the *helicola*, in which a spiral is wound around another spiral in both enlarging and diminishing directions. The helicola outlines the geometry of both micro- and macro-worlds and leads to a completely new understanding of multi-dimensionality.

2. Discovery of the *ultimate spiral field velocity* that is slightly greater than the velocity of light, and showing that the spiral field is formed as a result of deviation of its velocity of propagation from the ultimate spiral field velocity.

3

3. Discovery of two prime spiral elements of nature, *torix* and *helix*, that are particular cases of the helicola, with leading and trailing spirals representing electric and magnetic fields respectively.

4. Discovery of polarization processes that produce three types of torices from empty space by following the laws of conservation of energy and electric charge:

   Reality-polarized torices with mutually reverse flow of both energy, outward and inward, and time, forward and backward, that are respectively called the *real* and *imaginary torices*;

   Vorticity-polarized torices with mutually inverse direction of their vorticities, left-handed and right-handed;

   Complementary-polarized torices forming the particles of matter and dynamic ether.

5. Discovery of the relationships between the curvature and direction of vorticity of both torices and helices and their physical properties, including gravitational mass, inertial mass, and electric charge.

6. Discovery of two types of quantum energy levels of torices, *oscillation level* and *excitation level*, that create a great variety of both real and imaginary torices required to form elementary particles and *micro black holes*.

7. Discovery of the *unified nature of strong, gravitational and electric forces*, and explaining the creation of *nuclei* and *nucleon clusters*.

8. Discovery of a mechanism of formation of helices as a result of the shift of torices from higher to lower quantum energy levels, with the excitation helices forming photons and the oscillation helices forming neutrinos.

9. Rediscovery of the two principal types of waves predicted by Wheeler and Feynman's absorber theory that explained instantaneous transmission of information across the universe:

   Retarded waves, or conventional waves, that travel outwards from its source and forward in time, and associated with real torices;

   Advanced waves that travel backwards in time, from the future, to arrive at its source, and associated with imaginary torices.

10. Discovery of *dynamic ether* and a mechanism for its creation called the *Micro-Big-Bang inflation process.* The dynamic **ether** is the most likely candidate for the missing *dark matter.* It provides three principal functions:

   a)   Serving as a pool of embryonic matter, similar to the Higgs field, for the creation of new elementary particles;

   b)   Assisting in the transmission of electromagnetic waves and neutrinos, with a by-product in the form of the *microwave background radiation*;

   c)   Providing both the velocity and direction references to all the moving entities in the universe.

How This Idea Came to Be

This all started in 1992 from a discovery of a geometrical shape that perfectly represented the infinitude of the world and its repetitious character in both enlarging and diminishing directions. This shape turned out to be the endless multidimensional helical spiral which I called the *helicola.* In 1993, I sent my short essay "Multiple-Level Universe" to a number of scientists in the United States, and received several responses. The most encouraging of them was from Dr. Carlo Rovelli of the University of Pittsburgh who described my little essay as "delightful and stimulating." This was the beginning of the Spiral Field Theory (SFT) in which I expanded the concept of helicola into the micro-world of particle physics.

Nine years, two books and four papers later, I was at a crossroads. The main problem was that the SFT predicted the existence of two types of spiral fields forming matter. The properties of one type were described by real numbers that are commonly used in our every day experience. The properties of the other type, however, were described by imaginary numbers, or proportional to the square root of a negative number. Moreover, the spiral radii of this type of fields could be either positive, as one would normally expect, or, strangely negative. Since at that time, these exotic properties did not make any physical sense to me, I simply ignored the second type of spiral fields. This, however, limited my theory to a mere novel description of the known phenomena, without answering the principal questions related to the origin of matter and unification of the forces of nature.

The first breakthrough came about unexpectedly, after studying the concept of positive and negative energies. All of a sudden, the physical meaning of the imaginary spiral fields became very clear. Shortly after this discovery, I published a paper in which the elementary particles were presented as a product of polarization of a primordial space having zero energy and zero electric charge. The created particles were formed from the spiral fields with mutually reverse flow of energy and with opposite vorticities. Subsequently, the total energy and the electric charge of the particles remained equal to zero, or exactly the same as it was before their creation. When these particles changed their quantum energy levels, they either emitted or absorbed both the conventional electromagnetic waves that travel outwards from its source and forwards in time and the advanced waves that travel backwards in time. This was a remarkable confirmation of Wheeler and Feynman's absorber theory that explained instantaneous transmission of information across the universe.

The second, even more impressive, breakthrough came about after I used the SFT to derive the equations for the strong force between nucleons. As the reader probably knows, the strong force is currently considered as not related to any other forces of nature. My calculations, however, showed that the strong force have electrical nature. The long-awaited unification of the strong and electric forces became possible by considering the spiral structure of nucleons, confirming the powerful potential of the SFT.

Features of The Spiral Field Theory

This is the fourth book of a series that was set to describe the SFT, one of the most promising directions in development of a unified field theory. The first two books, *Spiral Grain of the Universe* and *Unified Spiral Field and Matter*, were written in the form of a story describing the contributions of many scientists to the theories based on the spiral concept. In the third book *Unified Spiral Nature of the Quantum & Relativistic Universe* and in this book, the reader will find a step-by-step explanation of the SFT that yields certain structures and physical properties of the fundamental components of nature, including nucleons, atoms, electromagnetic fields, and neutrinos. The theory is called *spiral field* because it considers the spiral fields, named the *torix* and the *helix*, as the prime elements of nature.

There are three reasons why the SFT may be helpful to those who are developing the so-called *unified field theory*. Firstly, it provides a path to the unification of two major theories, the theory of relativity and quantum mechanics, that are currently considered incompatible. Secondly, it leads to the unification of five principal forces: electric, magnetic, gravitational, strong, and weak. Thirdly, it provides some guidance in our search for the missing "dark" matter, which is believed to occupy from 90 to 99 percent of the universe.

The SFT has its roots in the vortex theories of 17th century, which were developed in application to the planetary system and later extended to atomic structure and electromagnetic waves. It strives to achieve the main goal of vortex theories, which is to produce a *visualizable* description of physical phenomena. Helping to accomplish this goal is an application of a special geometry of spiral fields that propagate along their spiral paths with the ultimate spiral field velocity. This approach provides the spiral fields with the characteristics that are closely related to the fundamental properties of matter, such as mass, electric charge, velocity, frequency, momentum, and spin. Subsequently, it becomes possible to explain how the torices form the atomic and nuclear particles, while the helices form the electromagnetic waves and neutrinos.

The reader may find some resemblances between the spiral fields of the SFT and the spiral elements used in other vortex theories, including Wheeler's geon, Penrose's twistors, Honig's photex, and Bostick's toroidal fiber. However, in spite of the apparent similarity with the forementioned spiral elements, the torix and helix have unusual characteristics that make the entire SFT unique. This brings us back to the fundamental question that has puzzled many scientists since the beginning of times. Where did the prime elements of nature come from? What sustains the continuous motion of electrons around the nuclei?

According to the SFT, at the very beginning there was empty space, or *nothing*, having zero energy and zero electric charge. The prime elements of nature were created by polarization processes that follow the law of conservation of energy and electric charge:

Total magnitudes of both energy and electric charge of the particles of the entire universe must remain equal to zero.

One shall have no problem with the electric charge remaining equal to zero. As kids, we were able to polarize an electrically neutral glass rod by rubbing it with silk. As grownups, we learned that the total electric charge of the system that includes both the glass rod and silk remains equal to zero. We may have, however, a problem with the total energy remaining at zero. What about the energy contained in the huge mass of the universe? Is it only the positive part of the total energy? If yes, then where is the remaining negative part that reduces the total energy to zero? One may find the answer to this question in the SFT. According to this theory, maintaining zero total energy is an important precondition for both the creation and the perpetuation of our universe. Dynamic equilibrium around zero spiral energy provides the stability for nucleons and atomic electrons. It makes atomic electrons run continuously around the nuclei without slowing down, like in a perpetual motion machine. It also makes the planets of our solar system do the same in respect to the Sun. It allows the electromagnetic fields and neutrinos to propagate across the universe with the constant velocity of light.

Understanding the role of positive and negative energies, however, is not easy, and requires us to overcome a deep-seated psychological barrier about the direction of the flow of time. We allow almost everything to be either positive or negative. This applies not only to physical properties, such as electric charge and temperature, but also to more ordinary things. We understand that a negative velocity means velocity in reverse direction, and a negative height of a hill is the same as the depth of a valley. When, however, it comes to time, our common-sense logic hinders us, and we stubbornly maintain that time can flow only in the positive direction. This is in spite of the fact that Maxwell's famous differential equations yield two solutions, one equivalent to a positive energy wave flowing into the future, and the other describing a negative energy wave flowing into the past.

You are not alone if you disagree with the part of the Maxwell's theory related to the time flow into the past. The majority of physicists also promptly forget this part soon after learning it at the university. This time, however, you may want to take another look at this phenomenon. By doing so, you will give yourself an opportunity to learn about a world that is based on a completely new principle. The world in which all the great ideas of the past become unified under a common umbrella, the world without unwelcome infinities forcing some of us to use forbidden mathematical tricks like renormalization. You will learn that all this becomes possible when the basic entities of the universe are presented in the form of *spiral fields* with special geometry.

It was a dream of many scientists to present a unified field theory with a limited number of equations that would fit on a half a page. My dream, however, was even more ambitious: to outline the theory with one sentence. To do that, I redefined the law of the conservation of energy and electric charge in the following manner:

Matter is a dynamic equilibrium of energy- and vorticity-polarized spiral fields at discrete energy levels.

So, these two short lines express the essence of the SFT. The rest is commentary.

Appealing Aspects of Spiral Field

There are three main reasons, in my opinion, why the concept of the spiral field appeals to the people of various backgrounds. The prime reason is our familiarity with the spiral shape that is found in various entities and processes of nature, and also in mythologies, legends, religions, and arts worldwide. In some cases, the presence of the spiral shape is most puzzling. We wonder why nature selected a double helical spiral to house the DNA? Why does the weightless air when twisted into a spiral shape of tornadoes and hurricanes, possess such gigantic destructive force? Why do celestial bodies and entire galaxies move along spiral paths? These, and many other puzzles, create a hunch that the role of the spiral shape in the universe is probably much greater than was previously thought.

The second reason is our desire to see an explanation of a theory in terms that we can readily visualize and comprehend. Contrarily to the majority of the current theories of physics that frequently employ advanced mathematical tools, the SFT theory uses the rather rudimentary math taught in high school. Also, instead of using some abstract multi-dimensional entities that are beyond our imagination, the SFT employs simple geometrical constructions to represent the basic elements of nature.

The third reason why the idea of spiral field is close to our hearts is our deep-seated intuitive gravitation towards things that are beautiful, smooth, dynamic, and periodic, with great varieties of exciting shapes, and a capability to be expanded endlessly in both enlarging and diminishing directions. These are, certainly, the attributes of the spiral fields.

You are welcome to this wonderful world of spirals. It is so immense that it leaves enough room for all of us: for those who may enjoy the ride by simply reading about spirals, and for those who may want to make their own contribution to the research on spirals, and also for those who may want to participate in this process by offering their honest and constructive criticism. There is, however, the danger of becoming one of us, the students of spirals. Once you have understood the concept of the spiral field, it will overwhelm you with its omnipresent logic, and you may begin looking at the scientific world from a completely different perspective. This path is only for brave people.

Acknowledgements

My acknowledgements go to all the scientists since the beginning of civilization whose works in any way affected the development of the SFT. Among my most admirable heroes are:

- The scientists of the world of Classical Antiquity. Among them are Anaximander of Miletus (611-547 B.C.), who first introduced the vortex concept to describe both the creation and the collapse of the universe, Democritus (470-400 B.C.), who considered the vortex motion as the general law of nature, Archimedes of Syracuse (287-212 B.C.) and Apollonius of Perga (c.260 -190 B.C.), who produced outstanding scientific works on spirals.

- Johannes Kepler (1571-1630), Renè Descartes (1596-1650), and Gottfried Leibniz (1646-1716), the principal contributors to the *theory of vortices* that considered the spiral as the grain of the Universe.

- Emanuel Swedenborg (1688-1772), who first applied the concept of vortices to the micro-world, attempting to explain various physical phenomena, including magnetism.

- Rudjer Bošković (1711-1787), who introduced the *universal force law* according to which the forces between the particles located at the close distances do not follow the inverse-square law and change their sign. Subsequently, he created a *unified theory* with an attempt to explain various physical and chemical properties of matter, including emission of light.

- Andrè Ampère (1775-1836) and Augustin Fresnel (1788-1827), who proposed the *theory of luminiferous molecule* that established a direct connection between the luminiferous ether and the atom.

- George Stokes (1819-1903), Hermann Helmholtz (1821-1894), Eugenio Beltrami (1835-1900), and Victor Schauberger (1885-1958), who enriched science with their studies of vortices in fluids.

- William Thompson, also known as Lord Kelvin (1824-1907), and Peter Tait (1831-1901), who developed the *theory of electromagnetic vortices* to explain properties of atoms and fields. Lord Kelvin deserves special praise for his loyalty to the idea of the vortex atom during the time when almost all scientists had abandoned it. In his 1889 speech, he predicted that the enormous difficulties of making this idea work will eventually be resolved.

- James Clerk Maxwell (1831-1879), the creator of the *theory of electromagnetism* that yields two solutions, one equivalent to the positive energy wave flowing into the future, and the other equivalent to the negative energy wave flowing into the past.

- Joseph John Thompson (1856-1940), who produced an excellent mathematical analysis of vortex rings in electromagnetic ether. This work led him to an investigation of cathode rays, and the subsequent discovery of the electron.

- Max Planck (1858-1947), the originator of one of the most fruitful theories of physics, *quantum mechanics.*

- Albert Einstein (1879-1955), well-known for his two extremely important propositions: (1) that the velocity of light is the maximum velocity of propagation for any object in the Universe and (2) that gravitation is caused by the curvature of empty space.

- Louis De Broglie (1892-1987), who stipulated that all forms of matter have both wave and particle properties. According to his theory, accompanying an electron is a wave (not an electromagnetic wave!) that guides the electron through space.

- Hannes Alfvèn (1908-1995), a pioneer in developing the *theory of magneto-hydrodynamic vortices in plasmas.*

- John Wheeler (b. 1911), who introduced in his *theory of geometro-dynamics* an entity called *geon*, that is "most visualized as a standing electromagnetic wave, or beam of light, bent into a closed circular toroid of high energy concentration."

- Richard Feynman (1918-1988), who jointly with John Wheeler, developed the *absorber theory*, explaining instantaneous transmission of light, based on the existence of two types of waves, the "retarded" waves in which time flows in a conventional way, and the "advanced" waves in which time flows in the opposite direction.

- Francis Crick (b. 1916) and James Watson (b. 1918), the co-discoverers of double-helical structure of DNA.

- Winston Bostick (b. 1916), who proposed the model of the electron in the form of the string-like submicroscopic force-free electromagnetic toroidal plasmoid.

- Peter Ware Higgs (b. 1929), who proposed the existence of the *Higgs field* that is responsible for providing elementary particles with mass.

- Roger Penrose (b. 1931), the creator of the *twistor theory* that presents an object called the *twistor*, which moves along in a spinning motion, as a building block of quantum space.

- Alan Harvey Guth (b. 1947), the author of the so-called *inflationary hypothesis* of cosmology, a substantial refinement to the basic *Big Bang theory*.

- Yoichiro Nambu, Holger Nielson, Leonard Susskind, Brian Greene and many other scientists, who developed the *string and supestring theories* that consider elementary particles to be vibrating and rotating one-dimensional strings propagating with the velocity of light.

Let me complete the list of the most distinguished personalities whose works influenced the development of the spiral field theory with the names of two outstanding people, Victor Schauberger (1885-1958) and Walter Russell (1871-1963). Both men, lacking formal education, achieved the highest level of understanding of many physical phenomena through their close connections with nature. They believed nature tells her secrets to those who listen. Both men became ardent proponents of the spiral grain of the universe.

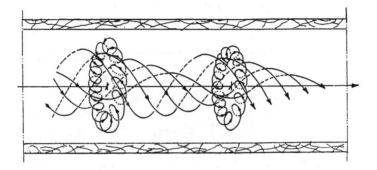

Fig. A1   Schauberger's presentation of flow of water inside of a wooden pipe with helical and toroidal vortices.

The following quotation from Walter Russell was provided by Glenn Clark in his enlightening book *The Man who Tapped the Secrets of the Universe*:

> That the man calls matter, or substance, has no existence whatsoever. So-called matter is but waves of the motion of light, electrically divided into opposed pairs, then electrically conditioned and patterned into what we call various substances of matter. Briefly put, matter is the motion of light, and motion is not substance. It only appears to be. Take motion away and there would not be even appearance of substance.

This statement keeps me shivering every time when I read it because of its unbelievably close forecast of the essence of the spiral field theory.

I am also thankful to my son Gene and daughter Ellen. In December 1992, Gene, who is now working in the computer field, has triggered my research on spirals by asking some very profound questions about the universe. He also nicely illustrated all my three books on the SFT and developed a first-class website on this subject. Ellen, who is a professional violinist, was also very helpful by editing my two books. I also enjoyed having several interesting discussions with her about spirals that certainly stimulated my thinking. The idea of a spherical spiral field actually belongs to her. My great thanks go to my brother Paul who took time to verify the derivations of the equations printed in this book, and also to my wife Tanya, who not only patiently put up with my devotion to the research on spirals, but also created favorable conditions for doing this work.

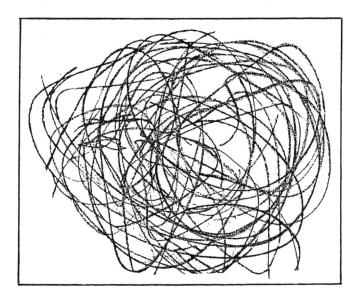

Fig. A2 A drawing of a spiral made by two-year-old Alexandra Orner.

I am especially grateful to my two-year-old granddaughter Alexandra, who gave me a wonderful present: several of her drawings of spirals. A copy of one of her spirals is shown above.

Thank you, all. I love you very much.

Vladimir B. Ginzburg

B

HISTORICAL BACKGROUND

The Roots of Spiral Field Theory

The roots of the spiral field theory can be found in the mythology and legends of prehistoric times. The explanation of the universe was profoundly influenced by impressive dynamic power of water and wind expressed in a spiral motion that came to be called the *vortex*. Thus, the whirlpool or water spiral is a part of world-wide folklore, and is a magical and religious symbol for the sources of life and energy.[1] A number of mythologies suggest that the constellation of the Great Bear, controlling the seasons, was kept revolving by a celestial whirlpool or whirlwind.

According to a summary of various myths made by Lugt,[2] life started in the water of the primeval vortex. The whirlpools were considered gates to the netherworlds where danger lurked. Whirlwinds provided paths for gods, demons, and witches. According to the Bible, God uses the whirlwinds as the mechanisms to come in contact with mortals:

> *Then the Lord answered Job out of the whirlwind.*
> *Job 38.1*
> *. . . and Elijah went up by a whirlwind into heaven.*
> *Kings (2) 2.11.*

Humans associated the revolution of the celestial bodies around the north star with the heavenly vortex that was also the source of all energy.[3] We find in the oldest sacred Hindu script known as the Revelation of Rigveda that the embryo lies in the depth of the primeval vortex. According to the tales of Zuni Indians, the World Mother created life by stirring water in an earthen bowl.[1] In Minoan culture the entrance to the underworld had a spiral shape, immortalized by the Cretan labyrinth of King Minos.[4]

The Polynesians thought that the whirlpool was the entrance to the world of the dead. According to a Finnish legend, this whirlpool goes all

15

the way through the earth.[5] The Maori of New Zealand believe that the souls of the dead go to heaven in a whirlwind.[1] Some Indian tribes dance along spiral paths because demons use such roads. Sumerians probably venerated a goddess of whirlpool by the name Is-hara of Ur who was connected with the sea serpent.[1]

In both Japan and China, the spiral is the symbol for the curled dragons. A Japanese serpent-god causes whirlwinds. The straits of Naruto in Japan, in which extremely large tidal vortices occur, are considered to be the "eastern gate of the dragon palace." According to a Chinese legend, a dragon-like serpent-god rules the universe in the form of a celestial vortex. In the Chinese Yin-Yang doctrine, the spiral is closely related to Yang, the source of life and energy.[6]

Vortex myths and legends are also popular in Celtic and Teutonic cultures. Here again, the whirlpool was the birthplace for life. In Scotland, even Christian religion adopted the pagan spiral motif for life. In Scandinavian mythology, the starry sky is rotated by the "world mill." Mythology of the spiral nature of the universe made a significant impact on the development of the later theories of both macro- and microcosm.

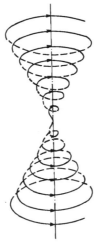

Fig. B1
Swedenborg's spiritual (upward) and natural (downward) minds.

The concept of the spiral was central to the prophetic teachings of the Swedish philosopher and physicist Swedenborg[7-9] (1688-1772), who proposed the existence of two minds, natural and spiritual (Fig. B1):

> The natural mind is curved into spirals from right to left, while the spiritual mind is curved from left to right, so that the two minds are

turned against each other, in reverse. This is the clue that devil dwells in the natural mind, and works from that base against the spiritual mind. Further, spiraling from right to left is downward and therefore toward hell, while spiraling from left to right moves upward, and therefore toward heaven.[9]

Ancient Philosophers

Greek philosophers envisioned the key concepts of the spiral field theory related to creation of matter. According to the Greek mathematician and astronomer Anaximander of Miletus (611-547 B.C.), our world came into being when a mass of material was separated from the Infinite by the rotary motion of a vortex. Subsequently, heavy materials concentrated at the center, while masses of fire surrounded by air went to the periphery and formed the heavenly bodies. Further, he held that all things eventually return to the element from which they originated.[10]

Another Greek philosophers Anaximenes of Miletus, also explained the formation of our world by the rotatory motion of primordial substance (Fig. B2). Similarly, Anaxagoras (500?-428 B.C.) proposed in his book *The Nature of Things* that all-pervading intelligence set the universe in motion by rotating the universal elements.

Fig. B2
Anaximenes's spiral
Universe.

About the same time, another Greek philosopher Leucippus further expanded the ideas of Anaxagoras, proposing that the vortex resulted from random collisions of atoms.[3] Atoms of irregular shape became entangled and subsequently moved towards the center. From this spiraling population of universal elements came the "stars and the Sun and the Moon and air." A similar theory appeared in the second century BC in China.[11, 12]

One of the strongest supporters of Leucippus's theory was his student Democritus (470-400 B.C.). Best known as an atomist, Democritus applied the vortex concept to explain the microcosm. According to his hypothesis, when a group of atoms becomes isolated, a whirl is produced which causes like atoms to tend toward each other. He considered the vortex motion so fundamental that he interpreted it simply as a general law of nature[2]. This was probably the first attempt in the history of science to formulate a "unified theory of physics."

Aristotle (384-322 B.C.) saw the analogy between the vortex proposed by these yearly philosophers and both the whirlpool and the whirlwind, "reasoning from what happens in liquids and in the air, where larger and heavier things always move towards the middle of the vortex.[2]" Two Greek philosophers became well-known for their fundamental research of the spiral shape. Archimedes of Syracuse (287-212 B.C.) made mathematical investigations of the two-dimensional spiral that carries his name[13]. He also invented new mechanisms and devices, and perfected some known ones that employed spiral motion. His younger contemporary, Apollonius of Perga (c.260-190 B.C.), worked out the mathematical aspects of a three-dimensional helical spiral.[10]

All these philosophical ideas and scientific studies of the ancient philosophers were put aside for almost eighteen hundreds years.

The Cartesians

Two major events of the 16th century, the Copernican revolution and discovery of magnetism by English physicist and physician William Gilbert (1544-1603), led to the revival of the vortex theory. In 1609, Johannes Kepler (1571-1630) proposed in his book *Astronomia Nova* that the Sun itself propagates a "species immateriata" similar to rays of light and arising from the Sun's magnetic force which turns with the Sun "after a manner of an impetuous vortex, which extends over the whole width of the world, and at the same time bears along the planets..." In Kepler's theory, the vortices were harmonic, providing musical relationships between the orbits of the planets.[14,15] These musical relationships, he thought, were given to humankind as a gift from heaven.

In the years 1629-33, René Descartes (1596-1650) independently proposed a similar concept that became known as the *vortex theory*[16]. He envisioned the matter in the sky in the form of vortices (Fig. B3). The planets were turning around the Sun, as a vortex with the Sun at its center. Similar to the whirlpools in the rivers, the parts of the vortex matter that were nearer to the Sun moved faster than those that were farther away.

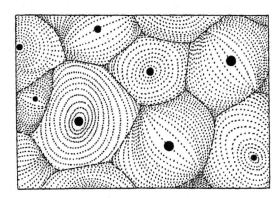

Fig. B3
Descartes' celestial
vortices.

The vortex theory have found numerous enthusiastic supporters and in the middle of 17th century, one hundred years after the birth of Isaac Newton, this theory was still considered by many distinguished scientists as viable alternative to Newtonian mechanics. Christiaan Huygens (1629-1695) strongly opposed Newton's use of attractive force as a fundamental explanatory principle. Force, in the Newtonian sense, could never count as a fundamental mechanical principle for Huygens. The occurrence of such forces always required a further, mechanistic explanation for him.

Alternatively, Huygens further advanced the vortex theory to explain gravity. He replaced cylindrical vortices posited by Descartes that could explain only the gravity toward the axis with a multilaterally moving vortex. In this vortex, the particles circle the earth in all directions, explaining a truly centrally directed gravity.[10]

Gottfried Leibniz (1646-1716) attempted to explain the elliptical orbits of the planets based on the vortex theory. The lack of such an explanation was considered by Newtonians to be a fundamental flaw of this theory. Leibniz supposed that each planet was involved in two motions, a trans-radial motion in which it moved exactly as the vortex fluid did and a radial motion in which it moved from layer to layer. Thus, the combined trajectory became elliptical.[17]

Emanuel Swedenborg (1688-1772) further extended the concept of vortices[7-9]. In his *Principia*, which was an obvious challenge to Newton's famous work, Swedenborg followed Democritus by proposing that the spiral motion is a fundamental principle, or "the first Simple" as he called it, of the Universe:

> . . .in a Simple, there is an internal condition tending to a spiral motion, and consequently there is in it a similar endeavor to produce it.[7]

Swedenborg used his theory to explain magnetism and even spiritual phenomena.

Leonhard Euler (1707-1783) aspired to create a uniform picture of the physical world. Although he was not a direct representative of Cartesianism, nevertheless he had been closer to Cartesian natural philosophy than to Newtonism.[10] He thought that the universe was filled up with ether - a thin elastic matter with extremely low density, like super-rarified air. The main property of ether particles is their impenetrability. To explain magnetism, Euler introduced magnetic whirls that are even thinner and move more quickly than ether.

The interest in vortex theory in France and the high esteem in which it was held is evidenced by the many essays based on the theory that were awarded prizes by the Paris and other Academies of Sciences up to 1740. However, despite the intelligibility of its qualitative mechanism, the vortex theory failed to predict quantitatively many phenomena. By the middle of the 18th century, it was abandoned by a majority of scientists in favor of Newtonian theory. A new understanding of vortex motion was a century away.

The downfall of the vortex theory was probably accelerated by the research of some scientists who were not comfortable with either vortex or Newtonian theories. The most distinguished among these scientists was Rudjier Bošković (1711-1787). He challenged both Leibniz and Newton with his Universal Force Law that predicted repulsive gravitational forces acting at close distances between particles.[18] He also published the first comprehensive study of tornadoes.[19]

Some scientists tried to find a compromise between the two theories. One of the most advanced hypotheses was proposed by Immanuel Kant in 1755 and Pierre Laplace in 1796. According to their hypothesis, the Sun and planets were developed out of a rotating gaseous cloud. Here, however, the idea of the vortex was introduced within the framework of Newtonian mechanics.

The Age of Electricity

The temporary downfall of the vortex theory in the 18th century did not stop creative thinking along its lines. The fruitful aspects of this theory became even more obvious with the advent of electricity. The idea of the existence of electromagnetic ether was proposed by André Amperé (1775-1836) and his close friend Augustin Fresnel (1788-1827) in their theory of the electrodynamic molecule[10].

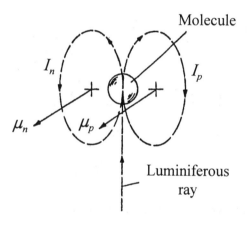

Molecule

Luminiferous
ray

Fig. B4
Ampére and Fresnel's
electrodynamic molecule.

They believed that the origin of the molecular currents was in the luminiferous ether that pervades space and penetrates matter (Fig. B4). The luminiferous ether was assumed to be electrically neutral. Inside the molecule of iron, the luminiferous ether decomposes into two separate electric liquids, positive I_p and negative I_n. Since these fluids are circling around the molecule in opposite directions, their molecular magnetic moments μ_p and μ_n are pointed in the same direction.

The rebirth of the vortex theory in the 19th century is a direct consequence of the works on the vortex motion in fluids by Euler (1707-1783), Stokes (1819-1903), Helmholtz (1821-1894), and Beltrami (1835-1899). Helmholtz saw a deep analogy between the magnetic field produced by electric current and the flow of incompressible fluid. He succeeded in correlating many theorems in electricity and hydrodynamics. His discovery of the permanency of vortex rings in a nonviscous fluid, convinced William Thomson, better known as Lord Kelvin (1824-1907), that atoms were vortex rings in such a perfect fluid, and that ether was a mixture of fluid elements.[20,21]

During his popular demonstrations before the Royal Society of Edinburgh, his long-time collaborator Peter Tait produced the smoke rings by mixing acid with ammonia[22]. These smoke rings bounced off each other, shaking violently from the collision, making the impression that they were made of rubber. William Thomson used this demonstration to point out that these vortex rings of smoke behaved just like atoms, and that all the properties of atoms stemmed from vortex spin. Drawing a parallel with the structure of the atoms made of toroidal vortices, he proposed that light waves were made of helical vortices (Fig. B5).

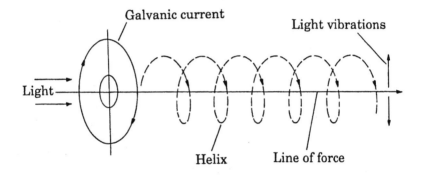

Fig. B5 Thomson's presentation of the light wave.

James Maxwell (1831-1879) also became an enthusiastic supporter of the vortex atom and ether and proposed several of his own mechanistic models of vortices related to electromagnetism[23,24]. There is no doubt that these vortex models were helpful in the formulation of his famous theory of electromagnetism in 1864. Unfortunately, because he presented his theory in terms of pure mathematics, many scientists have made erroneous conclusions about the role of mechanistic models. Since then, it became the custom among physicists to believe that mathematical equations alone represent physics. The vortex theory became one of the victims of this development.

In spite of this collateral damage to the vortex theory, Maxwell's theory will play a key role in the rebirth of the vortex theory in its current form. This is thanks to the fact that the Maxwell's differential equations yield two solutions, one equivalent to a positive energy wave flowing into the future, and the other describing a negative energy wave flowing into the past[25]. In the spiral field theory, this dual solution explains the creation of the reality-polarized spiral field[26].

The last prominent attempt to advance the theory of vortices in the 19th century was made by Joseph John (J. J.) Thomson (1856-1940). In 1882, in his Adams Prize winning *Treatise*[27], he proposed that atoms might be the vortex rings in the imagined electromagnetic ether. It was considered to be one of the best analytical research works on vortices produced to date. This work eventually stimulated his investigation of the cathode rays, that resulted in the discovery of the electron.

Like a century before, by the end of the 19th century, the vortex theory had again started to loose its best supporters. This time it yielded to two ambitious newcomers, quantum mechanics and the theory of relativity.

The Atomic and Nuclear Age

It took another century for the vortex theory to be born again. This time, however, it did not show up as a confrontational alternative to the widely accepted current theories. Contrarily, it embraced them as its own essential constituents. Moreover, it made compatible the quantum mechanics and the theory of relativity, a task which many scientists began to consider impossible. The product of this exercise was the spiral field theory, a subject of this book. Among the theories that mostly affected the development of the spiral field theory are:

Quantum mechanics - originally proposed by Max Planck (1858-1947) in 1900, this theory[28,29] states that the radiation is emitted or absorbed in energy packets called quanta. In the spiral field theory, the role of quanta is fulfilled by one of the prime elements of nature, the helices, that are either emitted or absorbed when the other prime elements of nature, the torices, change their discrete energy levels[30-33].

Special theory of relativity - Published by Albert Einstein (1879-1955) in 1905, this theory[34] considers the velocity of light as the maximum velocity in the universe. The spiral field theory[30-33] fully adopts the concept of the maximum velocity, and assigns this role to the ultimate spiral field velocity.

General theory of relativity - Published by Albert Einstein in 1915, this theory[34] explains gravity and, hence, the gravitational mass as a result of curvature of space-time (Fig. B6). Similarly, the spiral field theory[30-33] expresses the gravitational mass as a function of curvature of the dual spiral field.

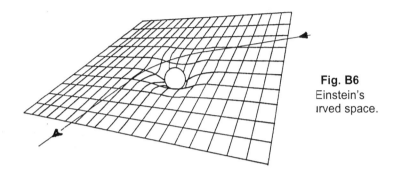

Fig. B6
Einstein's
ırved space.

Wave theory - Developed by Louis De Broglie (1892-1987) in 1923, it stipulates that "all forms of matter have both wave and particle properties." According to this theory[35-37], accompanying every electron is a wave (not an electromagnetic wave!) that guides the electrons through space, an idea that found its implementation in the spiral field theory in the form of the double toroidal spiral field[30-33] representing the atomic electron (Fig. B7).

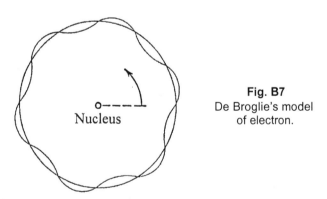

Nucleus

Fig. B7
De Broglie's model
of electron.

Absorber theory - According to this theory[25] developed by Richard Feynman and John Wheeler in the early 1940s, the charged particles radiate two types of waves, the "retarded" waves in which time flows in a conventional way, and the "advanced" waves in which time flows in opposite direction. Interference of the "retarded" and "advanced" waves produces instantaneous connections between the charged particles of the universe. In the spiral field, the "retarded" and "advanced" waves are created by reality-polarized helical spiral fields[26].

Geometrodynamics - Published by John Wheeler in 1962, this theory[38] introduces an entity called *geon* that "is most visualized as a standing electromagnetic wave, or beam of light, bent into a closed circular toroid of high energy concentration." Similarly, in the spiral field theory[30-33], the energy propagates along toroidal spiral path with the ultimate spiral field velocity.

Twistor theory - Introduced by Roger Penrose in early 1960's, this theory[39] presents a spinning and moving along object called the *twistor* to build up the notion of quantum space. The theory employs the algebra of complex numbers to explain many physical phenomena including electromagnetic waves and gravity. The spiral field theory also defines both its prime entities, the torix and helix, as the spinning and moving objects. Also, similarly to the twistor theory, the spiral field theory[30-33], employs the algebra of complex numbers.

Concept of fractional charges - Murray Gell-Mann and George Zweig[25,40] proposed in 1964 that the nuclear particles are made from quarks. Some quarks have 2/3 units of electric charge, and some have -1/3 units of electric charge (where the electron has -1 unit). Similarly, according to the spiral field theory, the nuclear particles are made of the torices with fractional electric charges[30-33]. Mills and Good proposed in 1995 that the electron in a hydrogen atom can be located at fractional quantum energy levels[54]. The spiral field theory applies a similar concept in respect to the torices forming the atomic electrons of a hydrogen atom[30-33].

String theory - proposed in 1970 by Yoichiro Nambu, this theory[40] states that elementary particles are vibrating and rotating multidimensional strings that propagate with the velocity of light. The string theory that incorporates supersymmetry became known as the *superstring theory*. It was developed by several research workers, including Green, Schwarz, and Dziarmaga[41,42]. Both the double torix and double helix of the spiral field theory can also be considered as particular cases of the strings.

Besides the theories mentioned above, there are numerous scientific works that may be related to the spiral field theory. Among them are:

- Models of spiral cosmic plasma[43-46]
- Models of elementary particles[47-56]
- Hydrodynamic models of vortices[57-61]
- Electrodynamic models of vortices[62-69]

- Models of force-free magnetic fields[70-72]
- Zero-point energy theory[73-82]
- Theories of superluminal velocity[83-98].

Universal Character of Spiral Shape

Numerous enlightening works on spiral geometry in nature were published in 20th century [2,57-60,99-114]. These works were most beautifully summarized in the book *The Seven Mysteries of Life* by Guy Murchie[102]:

If impalpable space thus turns out to be not nothingness but a mystic presence, we may ask of what it is made beyond the already mentioned sparsely strewn molecules of hydrogen, oxygen, carbon and dozens of life-harboring compounds. What is its structure? Does it have any definable grain of measurable texture that could be thought of as an integral part of it? . . . Yes, the firmament as we know it in our finitude of space-time has a grain. *Its grain is spiral* [bolding added] with a propensity toward abstract nodes or waists as in an hourglass. . . .

. . . Spirality pervades the Universe concretely and abstractly, visibly and invisibly. Not only do the galaxies of stars tend to this rotary form, and possibly the super-galaxies or even the Universe itself, but the spiroid theme dominates the courses and shapes of bodies and substances all the way down to the atom. The elliptical orbit of any moving body or particle inevitably spirals in time through space, uttering the world lines of the Sun, Moon, stars and planets. And fluids both liquid and gaseous, as we have seen, corkscrew their way ahead – the great currents in the ocean, rivers, whirlpools, hurricanes and jet streams in the sky and all the smaller storms from thunder cells and tornadoes down to dust devils in the desert, snowdrifts around a post and smoke plumes out of chimneys – even the celestial streams of plasma with their dark vortex spots on the Sun and stars.

The shapes of trees (*including family trees*), vines and most plants and their roots also tend to the helix, as do the forms of shells, certain crystals such as stalagmites, bones, pine cones, flowers, seeds, muscle fibers, worms, fungus mycelium and the molecules within them, particularly protein, long nerve cells and the twisted double-helix DNA in genes – and so on and on to the abstract lines of all genetic connections in evolution, gyrating magnetic lines, invisibly curved space and, more than likely, the still undefined cores of *atomic nuclei* [bolding added] – all eventually transcending into Urgrund, the space-timeless abstraction of Infinitude.

Unifying Character of the Spiral Field Theory

Spiral field theory has two major features that might be helpful in our search for a long sought *unified field theory*. The first feature relates to the unification of field, matter and ether which the spiral field theory holds to be merely various manifestations of the same spiral field. Subsequently, it possibly opens the door to unify under a common umbrella all the seemingly incompatible physical phenomena and entities of the universe, including:

- Matter and antimatter
- Ordinary and dark matter
- Electromagnetic fields and neutrinos
- Elementary particles and ether particles
- Electromagnetic, weak, strong, and gravitational forces.

The second feature of this theory relates to its inherent capability to unite people. Nowadays, the people of different races, religions, and nationalities begin to discover that the concept of spirals has existed in their legends, myths, traditions, and religions since the beginning of time. Moreover, the meaning of spirals turned out to be very similar for most of them in spite of the great diversity of their heritage and locations around the world. Hopefully, in the future, the concept of spirals will generate the sense of unity among the people of various social, economical, and educational backgrounds, also among the people having completely dissimilar personal interests. It would be the direct result of the widespread presence of the spiral shape in nature, and its inevitable presence in a great variety of branches of science and applied disciplines, including:

Archeology, architecture, art, astronomy, biochemistry, biology, botany, cosmology, chemistry, design, engineering, economics, genetics, history, hydrodynamics, instrumentation, mathematics, meditation, meteorology, music, mythology, philosophy, physics, religion, sociology, technology, etc.

Because the subject of spirals in nature is so immense and at the same time virtually unexplored, more and more people tend to become *students of spirals*.

REFERENCES

1. Mackenzie, D. A., *The Migration of Symbols and Their Relations to Beliefs and Customs*, A. A. Knopf, New York, 1926.
2. Lugt, H. J., *Vortex Flow in Nature and Technology*, Krieger Publishing Company, Malabar, Florida, 1995, p. 3-17.
3. Schuchhardt, *Alteurope*, 4th ed. Berlin, W. de Gruyter, 1941, p. 205.
4. Neumann, *The Great Mother*, Prinston University Press, Second Printing, 1974.
5. de Santillana, G. and von Dechend, H., *Hamlet's Mill*, Gambit, Ipswitch, Mass., 1969.
6. Purce, J., *The Mystic Spiral - Journey of the Soul*, Thames and Hudson, New York, 1997.
7. Swedenborg, E., *The Principia*, Swedenborg Institute, Basle, 1954.
8. Jonsson, I. *Emmanuel Swedenborg*, Twayne Publishers, Inc., New York, 1971.
9. Dole, F., *Emmanuel Swedenborg - A Continuous Vision*, Swedenborg Foundation, Inc., New York, 1988.
10. Gillispie, C.C., *Dictionary of Scientific Biography*, Charles Scribner's Sons, New York, 1971.
11. Aiton, E. J., *The Vortex Theory of Planetary Motions*, MacDonald, London and American Elsevier, Inc., New York, 1972, p. 34.
12. Needham, J., *Science and Civilization in China*, London, 1962, Volume 2, p. 371-372.
13. Dijksterhuis, E .J., *Archimedes*, Princeton University Press, Princeton, N.J., 1987.
14. Caspar, M., *Kepler*, Abelard-Schuman, London, 1959, p. 123-141.
15. Moore, P., *Watchers of the Stars*, G.P. Putnam's Sons, 1974.
16. Vrooman, J. R., *René Descartes*, G. P. Putnam's Sons, New York, 1970.
17. Aiton, E. J., *Leibniz*, Adam Higler Ltd., Bristol, 1985.
18. Boscovich, R. J., *A Theory of Natural Philosophy*, The M.I.T. Press, Cambridge, Mass., 1966.
19. Wegener, A., *Wind- und Wasserhosen*, Vieweg, Braunschweig, 1917.

20. Sharlin, H.I., *Lord Kelvin*, The Pennsylvania State University, University Park, Pa. and London, 1979.

21. Thomson, W., "Dynamic Illustration of Magnetic and Helicoidal Effects of Transparent Bodies on Polarized Light," *Proceedings of the Royal Society*, 7, p. 150-158, 1856.

22. Gray, A., *Lord Kelvin - The Dynamic Victorian*, Sharin, H. I., The Pennsylvania State University, University Park and London, 1979.

23. Domb, C., *Clerk Maxwell and Modern Science*, University of London, The Athlone Press, 1963.

24. Tricker, R. A. R., *The Contribution of Faraday and Maxwell to Electric Science*, Pergamon Press, Oxford, London, 1966.

25. Gribbin, John, *Q is For Quantum, An Encyclopedia of Particle Physics*, A Touchstone Book, Published by Simon & Schuster, New York, 2000.

26. Ginzburg, V.B. "Dynamic Aether," *Journal of New Energy*, Vol. 6, No. 1, 2001.

27. Thomson, J. J., *Motion of Vortex Rings*, MacMillan and Co., London, 1883, Bath, UK, 1996.

28. Plank, M. *Eight Lectures on Theoretical Physics*, Dover Publications, Inc., Mineaola, New York, 1998.

29. Heiborn, J. L., *The Dilemmas of An Upright Man*, University of California Press, Berkeley, 1986.

30. Ginzburg, V.B., "Toroidal Spiral Field Theory," *Speculations in Science and Technology*, Vol. 19, 1996.

31. Ginzburg, V.B., "Structure of Atoms and Fields," *Speculations in Science and Technology*, Vol. 20, 1997.

32. Ginzburg, V.B., "Double Helical and Double Toroidal Spiral Fields," *Speculations in Science and Technology*, Vol. 22, 1998.

33. Ginzburg, V.B., "Nuclear Implosion," *Journal of New Energy*, Vol. 3, No. 4, 1999.

34. Einstein, A., *Relativity – The Special and the General Theory*, Crown Publishers, Inc., New York, 1952.

35. de Broglie, L., *The Revolution in Physics*, The Noonday Press, New York, 1953.

36. de Broglie, L., *New Perspectives in Physics*, Basic Books, Inc. Publishers, New York, 1962.

37. Serway, R. A., *Physics for Scientists & Engineers with Modern Physics*, Saunders College Publishing, Philadelphia, Pa.,1995.

38. Wheeler, J. A., *Geometrodynamics*, Topics of Modern Physics, Vol. 1, Academic Press, New York, 1962.

39. Peat, D. F., *Superstrings and the Search for The Theory of Everything*, Contemporary Books, Chicago, Ill., 1988.

40. Millar, D., et al, *The Cambridge Dictionary of Scientists*, Cambridge University Press, 1996.

41. Green, B., *The Elegant Universe*, W. W. Norton & Company, New York and London, 1999.

42. Dziarmaga, J., "String Model for Short Range Interactions of Vortices," *Physical Review D*, Vol. 48, No. 8, p. 3807-17, 15 Oct. 1993.

43. Alfvén, H., *Cosmical Electrodynamics*, 2nd Edition, Clarendon Press, Oxford, 1963.

44. Lerner, E.J., *The Big Bang Never Happened*, Vantage Books, Division of Random House, Inc., New York, 1992.

45. Bostick, W. H. "Mass, Charge, and Current: The Essence and Morphology," *Physics Essays*, Vol. 4, No. 1, March 1991.

46. Stevens, C. B., "The Plasma Focus Fusion Device - Universal Machine of the Future," *21st Century Science and Technology,* Winter, 38-45, 1988.

47. Barness, T. G., "New Proton and Neutron Models," *Creation Research Society Quarterly*, Volume 17, p. 42-47, June 1980.

48. Bergman, D. L. and Wisley, "Spinning Charged Ring Model of Electron Yielding Anomalous Magnetic Moment," *Galilean Electro-dynamics*, Vol. 1, No.5, Sept/Oct. 1990.

49. Brown, L. S. and Gabrielse, G., "The Geonium Theory: Physics of a Single Electron or Ion in a Penning Trap," *Review of Modern Physics,* Vol. 58, No. 1, p.233- 311, Jan.1986.

50. Ekstrom, P. and Wineland, D., "The Isolated Electron," *Scientific American*, Vol. 243, p. 104-118, Aug. 1980.

51. Gabrielse, G., "The Extremely Cold Anti-protons," *Scientific American*, Vol. 267, Dec. 1992, p. 78-89.

52. Lockyer, T. N., *Vector Particle Physics*, BookCrafter, Chelsea, MI, 1991.

53. MacGregor, M. H., "Evidence for Two-Dimensional Ising Sructure in Atomic Nuclei," *Il Nuevo Cimento*, Vol. 36A, No.2, p. 113-168, 21 Nov. 1976.

54. Mills, R. L. and Good, W. R., "Fractional Quantum Energy Levels of Hydrogen," *Fusion Technology*, 28, p. 1697-1719, 1995.

55. Rocher, E. "Noumenon: Elementary Entity of a New Mechanics,"

Jour. Math. Physics., 13(2), 1919-1925, 1972.

56. Wasserman, J., "On the Structure of the Electron," *Speculations in Science and Technology*, Vol. 15, 1992.

57. Alexandersson, O., *Living Water*, Gateway Books, Bath, UK, 1996.

58. Coats, C., *Living Energies*, Gateway Books, Bath, UK, 1996.

59. Ballabh, R., "On the Coincidence of Vortex and Streamlines in Ideal Liquids," *Ganita*, I(2), p. 1-4, 1948.

60. Bjorgum O., "On Beltrami Vector Fields and Flows (Parts 1 and 2)," Universitet I Bergen Arbok, 1951, 1952.

61. Purce, J., *The Mystic Spiral*, Thames and Hudson, New York, 1980.

62. Akimov, A. E. and Tarasenko, V. Y., "Models of Polarized States of the Physical Vacuum," *Izvestia Vysshikh Uchebnykh Zavedenii, Fizika*, No. 3, p. 13-23, March 1992.

63. Akimov, A. E. and Shipov, G. I., "Torsional Fields and Their Experimental Manifestations," *Proc. of the International Conf. on New Ideas in Natural Sciences*, St. Petersburg, Russia, June 1996.

64. Bass, R., "Self-Sustain Non-Hertzian Longitudinal Wave Oscillations," *Proc. of the Tesla Center Symp.*, 89-90, 1984.

65. de Aquino, F., "Gravitation and Electromagnetism; Correlation and Grand Unification," *Journal of New Energy*, Vol. 5, No. 2, 2000, p.76-84.

66. Bjorgum O. and Godal, T., *On Beltrami Vector Fields and Flows (Part 2)*, Universitet I Bergen Arbok, 1952.

67. Herzberg, G., *Atomic Spectra and Atomic Structure,* Dover Publication, New York, 1944.

68. Lakhtakia, A. and Weighofer, W. S., "Covariances and Invariances of the Beltrami-Maxwell Postulates," *IEEE Proc. – Science, Measurement and Technology,* Vol. 142 No. 3, p. 262-266, 1995.

69. Reed, D., "Novel Electromagnetic Concepts and Implications for New Physics Paradigms and Energy Technologies," *Journal of New Energy*, Vol. 2, No.1, Spring 1997.

70. Chandrasekhar, S., "On Force-Free Magnetic Fields," *Proc. Nat. Acad. Sci.* **42**(1), p. 1-5, 1956.

71. Chandrasekhar, S. and Waltjer, L, "On Force-Free Magnetic Fields," *Proc. Nat. Acad. Sci.* 44(4), p. 285-289, 1958.

72. de San, M. G., "An "Explicit" Solution of Force-Free Magnetic Fields?" *Speculations in Science and Technology*, Vol. 7, 1994.

73. Aitchison, I. J. R., "Nothing's Plenty: the Vacuum in Modern

Theory," *Contemp. Phys.,* **26**(4), 333-391, 1985.

74. Barness, T. G., *Space Medium: The Key to Unified Physics*, available from CRS Books, 1986.
75. Genz, H., *Nothingness - The Science of Empty Space*, Helix Books, Perseus Books, Reading, Massachusetts, 1994.
76. Kelly, D., "A Review of the Free Energy Scenario," *Speculations in Science and Technology*, Vol. 13(4), 1991.
77. King, M. B., *Tapping The Zero-Point Energy*, Paraclete Publishing, Provo, UT, 1989.
78. King, M. B., "Fundamentals of a Zero-Point Energy Technology." In Albertson, M. (ed), *Proc. of the Int. Symp. on New Energy*, p. 202-217, Rocky Mountain Research Institute, Fort Collins, CO, 1993.
79. King, M. B., "Dual Vortex Forms: The Key To a Large Zero-Point Energy Coherence," *Journal of New Energy*, Vol. 5, No. 2, 2000, p.25-37.
80. Michrowski, A., "Vacuum Energy Development," In Albertson, M. (ed), *Proc. of the Int. Simp. on New Energy*, p. 202-217, Rocky Mountain Research Institute, Fort Collins, CO, 1993.
81. Puthoff, H, "The Energetic Vacuum: Implication for Energy Research," *Speculations in Science and Technology*, Vol. 13(4), 1991.
82. Samokhin, A., "Vacuum Energy - A Breakthrough?" *Speculations in Science and Technology*, Vol. 13(4), 1991.
83. Avetissian, H. K. et al, "Superluminal Comton Laser," *Physics Letters A*, Vol. 137, No. 9, p. 463-465, 5 June 1989.
84. Cohen, H. H., et al, "Radio Sources With Superluminal Velocities," *Nature* 268, p. 405-409, 4 Aug. 1977.
85. Enders, A., and Nimtz, G., "On Superluminal Barrier Traversal," *J. Phys. I France 2*, p.1693-1698, Sept. 1992.
86. Feinberg, G., "Possibility of Faster-Than-Light Particles," *Physical Review*, Vol. 159, p. 189, 1105, July 1967.
87. Fox, R., et al, "Faster-Than-Light Group Velocities and Casualty Violation," *Proc. Roy. Soc.* London, A.316, p. 515-524, 1970.
88. Freedman, D.H., "Faster Than Speeding Photon," *Discover*, Vol. 19, No. 8, Aug. 1998.
89. Herbert, N., "Faster then Light: Superluminal Loopholes in Physics," *New American Library*, 1988.
90. Koryu, I. and Giakos, G. C., "Transmit Radio Messages Faster Than Light," *Microwaves and RF*, Vol. 30, p. 114-119, Aug.1999.
91. Milnes, H. W., "Faster than Light?" *Radio Electronics*, Vol. 54,

No.1, Jan. 1983.

92. Mirabel, I. F. and Rodrigues, L. F., "A Superluminal Source in the Galaxy," *Nature*, Vol. 371, p. 46-48, 1 Sept. 1994.

93. Pappas, P. T. and Obolensky, A. G. "Faster-Than-Light Claim Under Fire," *New Scientist*, p. 35, 14 Jan. 1989.

94. Petry, W. W., "Superluminal Velocity and Deviation from Newtons's Inverse Square Law," *Astrophysics and Space Science* 140, p. 407-419, 1988.

95. Phipps, T. E., "Superluminal Velocities: Evidence for New Kinematics?" *Physics Essays*, Vol 2, No.2, p. 180-185, 1989.

96. Rosen, A. "BL Lacertae Variability and Superluminal Motion via a Helical Filament- Shock Interface," *Astrophysical Journal* 359, p. 296-301, 20 Aug. 1990.

97. Sheldon, E., "Superluminal Motion," *Heavy Ions in Nuclear and Atomic Physics, Proc. of the 20th Mikolajki Summer School on Nuclear Physics*, Mikolajki, Poland, 1-10, Sept. 1988.

98. Squires, E. J., "Explicit Collapse and Superluminal Signals," *Physics Letters A* 163, p. 356-358, 1992.

99. Cook, T. A., *The Curves of Life*, Dover Publications, Inc., New York, 1979.

100. Ghyka, M., *The Geometry of Art and Life*, Dover Publications, Inc., New York, 1977.

101. Hargittai, I. and Pickover, C. A., *Spiral Symmetry*, World Scientific, London, 1993.

102. Murchie, G., *The Seven Mysteries of Life*, Houghton Mifflin, Boston, Mass., 1978.

103. Pickover, C. A., *Computers and the Imagination*, St. Martin's Press, New York, 1991.

104. Tomkeieff, S. I., *A Periodic Table of the Elements*, Chapman & Hall Ltd., London, 1954.

105. Watson, J. D., *The Double Helix*, W.W. Norton & Company, New York, 1980.

106. Russell, W., *A New Concept of the Universe*, The University of Science and Philosophy, Swannanoa, Waynesboro, VA, 1989.

107. Russell, W., *The Secret of Light*, The University of Science and Philosophy, Swannanoa, Waynesboro, VA, 1994.

108. Winter, D., et al, *Alphabet of the Heart-Sacred Geometry: The Genesis in Principle of Language and Feeling*, Eden, New York, 1994.

109. Carter, J., *The Other Theory of Physics - A Non-Field Unified Theory of Matter and Motion*, Absolute Motion Press, 2000.
110. Aquarian, A., *The Little Scroll*, P.E.A.C.E Publications ehf, Iceland, 2001.
111. Ginzburg, V. B. *Spiral Grain of the Universe - In Search of the Archimedes File*, University Editions, Huntington, WV., 1996.
112. Ginzburg, V. B., *Unified Spiral Field and Matter - A Story of a Great Discovery*, Helicola Press, Pittsburgh, PA, 1999.
113. Ginzburg, V. B., "Nuclear Implosion," *Journal of New Energy*, Vol. 3, No. 1, 1999.
114. Wolfram, S., *A New Kind of Science*, Wolfram Media, Inc. Champaign, IL, 2002.

C

HELICOLA

The First Spiral Principle

The *First Spiral Principle* was proposed by the author in 1993. It was initially formulated as an abstract geometrical concept [1], not related to any particular theory or model:

Every line is a spiral.

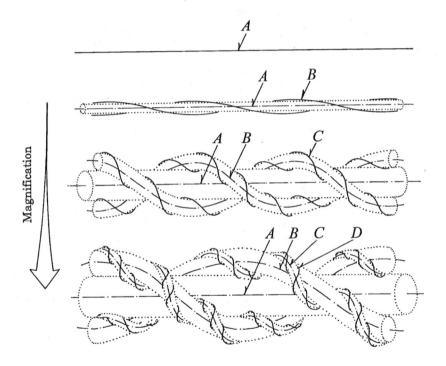

Fig. C1 The helicola.

What this means is that what we see as a line turns out to be, upon closer examination, a helical spiral. Although this principle may sound unjustifiable, there is actually nothing unusual about it. When you look at a thread with naked eye, it will appear to you as a line. But if you take a closer look at the thread with a magnifying glass, you may find it is made of even thinner treads spiraling around each other. We see the thread as a line because of the limitations of our vision. So, whatever appears as a line to us may actually be a spiral to beings of a much smaller scale. But the opposite situation will also be true. What we see as a long helical spiral will be seen as a line by beings of a much larger scale simply because of the finite resolution of their vision.

By following the First Spiral Principle, one will be able to discover the multi-dimensional spiral called the *helicola* (Fig. C1). First, draw a straight line A. Then apply the First Spiral Principle to discover that upon a close examination, that the line A actually looks like a helical spiral B wound around the line A. But the windings of the helical spiral B are also made of a line. Therefore, according to the first spiral principle, this line, under closer analysis, reveals itself to be a helical spiral C. Thus, the spiral C is wound around the spiral B. Is it the end of the story? Of course not, because the winding of spiral C is also made of a line which according to the first spiral principle appears as a helical spiral D wound around spiral C, and so on and on.

Helicola of Odd and Even Dimensions

As an example, let us first apply the First Spiral Principle to the construction of the helicola of the first three odd dimensions[2] (Fig. C2). The helicola of the first dimension is represented by line A with the radius of curvature $R_1 \to \infty$ and the spiral wavelength $\lambda_1 \to \infty$. It requires only one parameter x to define the position of a point m on the line. Since every line is a spiral, the one-dimensional line A becomes a three-dimensional helical spiral B wound around line A. In that case, three parameters are needed, the spiral radius R_2, the spiral wavelength λ_2, and the angle α, to define the position of the point m on the spiral B.

By further applying the first spiral principle, we may convert the three-dimensional helical spiral B into the five-dimensional spiral C wound around the spiral B that, in its turn, is wound around the straight line A. The five parameters that define the position of the point m on the spiral C are the spiral radii R_2 and R_3, the spiral wavelengths λ_3 and λ_4, and the angle α.

Even dimensions of the helicola are shown in Fig. C3. They start from the circular line *A* that represents the second dimension, requiring two parameters, the spiral radius R_1 and the angle α, to define the position of the point *m* on the circle *A*. One may think of the circle as a particular case of the helical spiral with the wavelength $\lambda_2 = 0$.

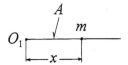

The first dimension

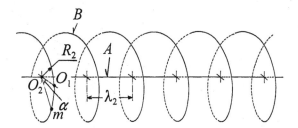

The third dimension

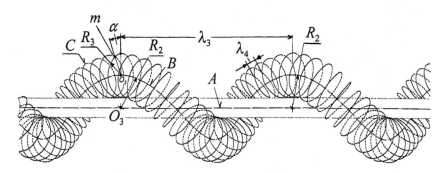

The fifth dimension

Fig. C2 Helicola of odd dimensions

The fourth dimension is represented by the spiral *B* wound around the circle *A* with four dimensions defining the position of the point *m* on the spiral *B*. These dimensions include the spiral radii R_1, and R_2, the spiral wavelength λ_2, and the angle α. The sixth dimension is represented by the spiral *C* wound around spiral *B* that, in its turn, is wound around spiral *A*. The six dimensions defining the position of the point *m* on the spiral *C* are the spiral radii R_1, R_2, and R_3, the spiral wavelengths λ_2 and λ_3, and the angle α.

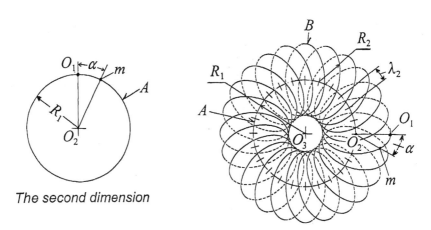

The second dimension

The fourth dimension

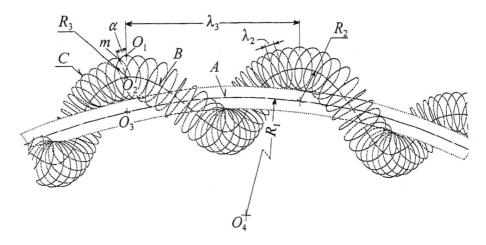

The sixth dimension

Fig. C3 Helicola of even dimensions.

Helicola of Celestial Mechanics

We will show in this book that the helicola is not merely an abstract geometrical shape, but the extraordinary shape that played a fundamental role in the construction of both the micro- and macro-worlds. To illustrate the point, let us briefly review how these ideas are applied to the celestial mechanics of our universe (Fig. C4, *a-c*).

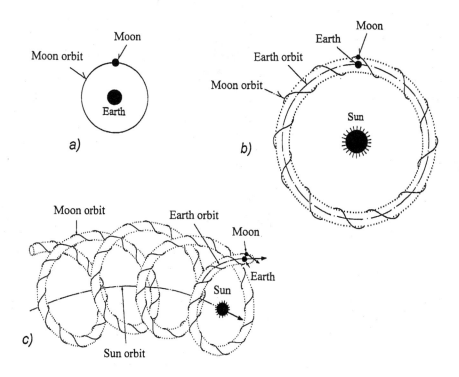

Fig. C4 Helicola of celestial mechanics.

To an observer standing at the center of the Earth, the Moon's path will appear as a circle (an ellipse to be precise) with the Earth in its center (*a*). However, the observer located in the center of the Sun will see the Moon's path as a toroidal spiral wound around a circular path of the Earth, with the Sun in the center of this circle (*b*). We will readily create another level of the helicola by relocating the observer to the center of the Milky Way. Now, the Moon's path will appear as a helical spiral wound around the helical spiral path of the Earth that is wound, in its turn, around a circular path of the Sun with its center at the center of our galaxy (*c*).

By extrapolating this concept further, one may conclude that the levels of the helicola continuously increase, making the entire universe a super-giant infinitely-expanding multiple-level helicola.

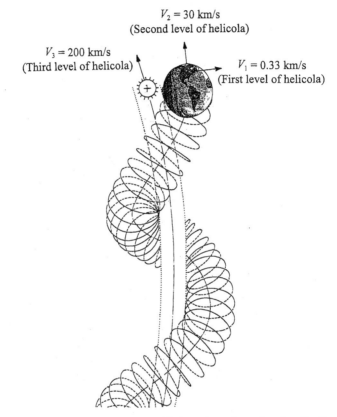

$V_2 = 30$ km/s
(Second level of helicola)

$V_3 = 200$ km/s
(Third level of helicola)

$V_1 = 0.33$ km/s
(First level of helicola)

Fig. C5 Spiral motion of a New Yorker along three levels of helicola.

All of us are moving continuously along these levels of helicola. For instance, as shown in Fig. C5, every five minutes a person living in New York will advance about 100 km along the first level of helicola that coincides with the trajectory of the rotation of the Earth's surface around its axis. During the same time, this person will also advance about 9000 km along the second level of helicola that coincides with the orbital path of the Earth around the Sun. In addition, during the same five minutes this person will advance 60,000 km along the third level of helicola that is the orbital path of the Sun around the center of our galaxy.

Obviously, the Milky Way is not standing still either. So, this person may have also advanced many hundreds of thousands of kilometers along the fourth level of helicola, also along the fifth, sixth, and so on and so on. This vision makes us believe that the spiral is at the core of the structure of the universe.

Current astronomical data allows us to define the dimensions of the helicola that represent the trajectories of four celestial entities, the Moon, the Earth, the Sun, and the Milky Way:

- The first dimension of helicola will be defined by the spiral trajectory of the Moon around the Earth
- The second dimension of helicola will be defined by the spiral trajectory of the Earth around the Sun
- The third dimension of helicola will be defined by the spiral trajectory of the Sun around its main orbit inside the Milky Way
- The fourth dimension of helicola will be defined by the main spiral trajectory of the Sun around the center of the Milky Way
- The fifth dimension of helicola will defined by the spiral trajectory of the center of the Milky Way around a still undetermined object of great gravitational pull.

Fundamental Elements of Nature

Spiral field theory addresses the problem that has confounded science since antiquity. What are the fundamental elements of nature? We have come forward a long way learning about atoms, electrons, protons, neutrons, quarks, and elementary particles. We originally perceived the fundamental element of nature as something solid and indestructible (remember that in Greek the *atomos* literally means "not to be cut"). We know today that this definition is no longer valid and we talk today about the fundamental elements of nature in terms of wave functions and vibrating strings.

We also originally thought of nature as a conglomerate of atoms in empty space. Many of us remember the famous quotation from the ancient Greek philosopher Democritus of Alberta: "Nothing exists except atoms and empty space; everything else is opinion." Since his time, we have learned a great deal about magnetic and electric fields. We also revived the idea of ether filling the empty space of the universe. But this is not all.

Some scientists believe that the space previously thought to be empty is filled with countless invisible particles containing unbelievable source of energy called the *zero-point* energy. Moreover, there also must be some room for the recently discovered *dark matter*, which, by some estimates, represents from 95 to 99% of the entire matter in the universe.

This tremendous variety of matter in the universe makes a task of discovering the unified fundamental elements of nature even more difficult. Our progress in this direction was so far very limited, although there are some hopes that *superstring theory* may lead someday to a breakthrough. Meanwhile, the spiral field theory offers another avenue that should be seriously explored. According to this theory, the fundamental elements of nature are the *torix* and the *helix* that are the smallest components of the double toroidal and double helical spiral fields. Geometrically, these fields are merely particular cases of the helicola, as will be described in the following chapters.

REFERENCES

1. Ginzburg, V. B., *Multiple-Level Universe*, IRMC, Pittsburgh, PA, 1993.
2. Ginzburg, V. B., "Unified Spiral Field, Matter and Ether - An Introduction to Spiral Field Theory," Conference *"Storrs 2000"*, University of Connecticut, June 4 - 9, 2000.

D

THE BASIC MATHEMATICS OF SFT

Math and Mechanical Models

Math is certainly indispensable for science, and particularly for physics. When a physical idea is expressed in mathematical terms, it becomes possible to analyze whether this idea confirms known experimental data. Also, it provides a very convenient tool to predict quantitatively new phenomena. However, math is not the only tool that helps us solve the problems of physics. Before the end of the 19th century, the majority of scientists believed that using math alone was not sufficient to create a comprehensible model.

Therefore, they developed mechanical models in which math went hand in hand with a common-sense description of physical phenomena based on known mechanical terms, so that one could readily visualize the physical concept. Bohr's model of the hydrogen atom was probably one of the best examples of mechanical models known to us. In his model[1] developed in 1913, Bohr presented hydrogen atomic structure as a miniature solar planetary system, with the nucleus representing the Sun, and the electrons representing the planets. Mechanical models were helpful for communicating ideas between scientists. They also helped the general public, not familiar with math, to understand the essence of these ideas.

The attitude towards the mechanical models began to change at the end of the 19th century. This change was actually triggered by one of the most important events in the history of physics, the discovery of electromagnetism by James Maxwell[2]. In 1864, after numerous unsuccessful attempts to describe the process of the propagation of light in mechanical terms, James Maxwell presented his theory in pure mathematical form. The triumphal experimental confirmation of this theory by Heinrich Hertz in 1888 was the beginning of a new trend in physics, when the majority of scientists adopted the position that math alone fully describes physical phenomena. This trend was epitomized by the frequently repeated quote from Hertz[3]. In answer to the question "What is Maxwell's theory?" Hertz replied, "Maxwell's theory is Maxwell's system of equations. . ."

This development produced two mutually opposite effects. On the one hand, it triggered a great interest in math among physicists and created the conditions for the development of several mathematically-intensive theories, such as quantum mechanics and the general theory of relativity. On the other hand, it separated physicists not only from the general public, but frequently even from each other. Albert Einstein used to complain that after physicists began to apply advanced math to his theory he himself no longer understood it. Unfortunately, the trend of idolizing the role of math continued throughout almost the entire 20th century, making it inconceivable for a person not versed in the most advanced math to become a theoretical physicist. By the end of the 20th century, developed theories of physics became so mathematically intense that some physicists bragged that to solve their equations they had to run the fastest computers in the world for several days.

A reasonable question to ask is whether this development in physics is justified. Is it really necessary to go that far into advanced math to describe nature, or is this merely an indication of having made wrong assumptions about the basic principles that govern the universe? Spiral field theory offers an alternative direction of research: It employs many fruitful traditions of the old times when math was only a part of the mechanical models, understood by both scientists and the general public interested in science. One may even suggest that the closer we get to the primordial level of matter, the simpler must be the math required for its description. This idea could be reduced to an extremely controversial proposition:

> *The basic concept of elementary matter shall be described by elementary math.*

The above proposition parallels the development of nature from a simple entity to a complex one. Math also develops from elementary to advanced. Any disproportional complexity of math is counterproductive, and is most likely caused by the inadequacy of the mechanical models employed.

Basic Equation of Spiral Field Theory

The math of SFT is based on the Theorem of Pythagoras[4] that states:

> *The square of the measure of the hypotenuse of a right triangle is equal to the sum of the squares of the measures of the legs.*

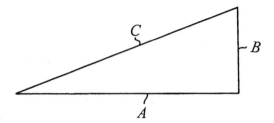

Fig. D1 Right triangle.

Let us review two cases. In the first case, we will consider a right triangle (Fig. D1) in which the measure of the hypotenuse is equal to *C*, and the measures of the legs are equal to *A* and *B*. Thus, in respect to *C*, the Theorem of Pythagoras is mathematically expressed by the equation:

$$C = \sqrt{A^2 + B^2} \qquad \text{(D1)}$$

Certainly, many of us learned this simple equation D1 when we were teenagers. However, recollecting it now will help us to manage fifty percent of the math required by SFT. Consider a numerical example. Let the measures of two legs of the right triangle to be as follows: *A* = 3 and *B* = 4. Then, after substituting these values into Eq. (D1), we find that the measure of the hypotenuse will be equal to:

$$C = \sqrt{3^2 + 4^2} = 5$$

If you feel comfortable with the above example, it means that you have successfully mastered the first half of the math required by the spiral field theory. Now, encouraged by this success, you are ready to embrace the slightly more difficult but still manageable second part of the math that outlines the spiral field theory.

The second part of the math still deals with the same Eq. (D1) based on the Theorem of Pythagoras. However, it imposes one important condition for the right triangle: the measure of at least one of the legs of the right triangle is greater than the measure of the hypotenuse.

Mathematically, this condition is expressed by the inequalities:

$$A > C \quad \vee \quad B > C \tag{D2}$$

As an example, let us consider a right triangle having the measure of hypotenuse $C = 4$ and the measure of one of the legs $A = 5$. To find the measure B of another leg of the triangle, we will rearrange the Eq. (D1) into the form:

$$B = \sqrt{C^2 - A^2} \tag{D3}$$

Then, after substituting the values of C and A into Eq. (D3), we obtain that the measure B is equal to:

$$B = \sqrt{4^2 - 5^2} = \sqrt{-9} = 3\sqrt{-1} = 3i$$

Contrarily to the first example, when all three measures A, B, and C were expressed with the *real* numbers, the measure B in the second example is expressed with an *imaginary number*, or a number that is presented as a square root of a negative number. As the reader will see from the following sections of this book, the imaginary numbers have deep physical meaning and describe the parameters of entities that play a vital role in the universe.

The Mechanical Model of a Single Helical Spiral

Let us start from the creation of a simple mechanical model called the *helical spiral*. As we mentioned before, mechanical model requires the phenomenon to be clearly described by using known mechanical terms. In this particular case, these terms will be lengths and points.

Consider a cylindrical can (Fig. D2a) with a diameter $d = 75$ mm and the height $\lambda = 100$ mm. Cut a rectangular piece of paper (Fig. D2b) with the length of one side equal to the perimeter of the can $\pi d = 3.14 \times 75 = 235.5$ mm and the length of the other side equal to the height of the can $\lambda = 100$ mm. After drawing the diagonal ab between the corner points a and b of the rectangular piece of paper, wrap it around and stick with either glue or a scotch tape to the can body. The diagonal *ab* will appear on the can as one wrap of a helical spiral (Fig. D2c). Geometry of this spiral is described by two parameters, the spiral diameter $d = 75$ mm and the wavelength $\lambda = 100$ mm.

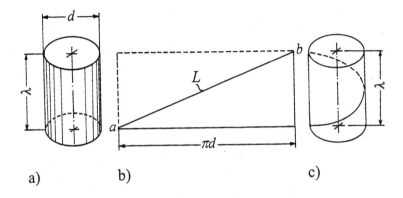

a)   b)   c)

Fig. D2 Construction of a single helical spiral.

Thus, one wrap of a helical spiral can readily be presented with a right triangle with the length of one of its legs equal to the spiral radial perimeter πd, and with the length of its other leg equal to the spiral wavelength λ. The length of the helical spiral L will be equal to the length of the diagonal *ab*. Therefore, according to the Theorem of Pythagoras, it is equal to:

$$L = \sqrt{(\pi d)^2 + \lambda^2} \tag{D4}$$

By constructing a helical spiral and establishing a mathematical relationship between the elements of the helical spiral given by Eq. (D4), we developed the mechanical model of the helical spiral. We are ready now to develop the mechanical models of two prime elements of nature, the *torix* and the *helix*. From a geometrical point of view, the torix and the helix are merely particular cases of the helicola.

The Torix

Geometrically, the torix is based on the helicola of infinite even dimensions. For the sake of simplicity, we showed in Fig. D3 a torix based on the helicola of the fourth dimension. This torix appears as a double toroidal spiral[5]. Construction of this spiral involves engaging a rod *ab* in two simultaneous motions, rotational and translational. The rod *ab* rotates with rotational velocity V_{2r} around its middle point. For an observer located at the middle point, the end points of the rod *a* and *b* will move along a circle with radius r_2. During translational motion, the rod's middle point advances with translational velocity V_{2t} along a base leading circle A_1 with radius r_1. The difference between the radii r_1 and r_2 is called *inversion radius r_i*.

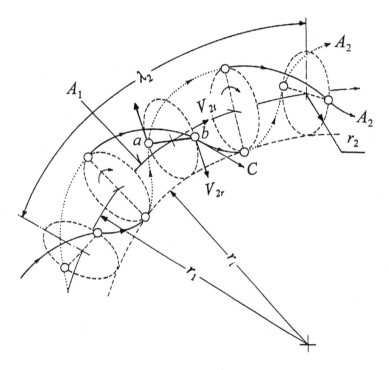

Fig. D3 Construction of a torix.

As a result of the simultaneous rotational and translational motions of the rod *ab*, the end points of the rod will move along two trailing toroidal spirals A_2. These two trailing spirals are angularly separated one from another by 180 degrees and form the trailing double toroidal spiral. The torix contains only one winding of the double toroidal spiral. After the torix completes one full revolution around the leading circle with the radius

r_1, it will leave a trace that is called the *torus* (Fig. D4).

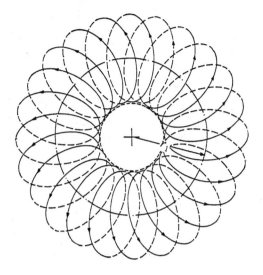

Fig. D4 Torus.

There are two principal geometrical features of the torix that enable it to become one of the two prime elements of nature:

1. Length of one winding of the trailing spiral L_2 is equal to the length of the leading circle L_1:

$$L_2 = L_1 = 2\pi r_1 \tag{D5}$$

2. For a group of torices that belong to the same family, the inversion radius r_i remains constant:

$$r_i = r_1 - r_2 = const. \tag{D6}$$

An additional feature of the torix relates to its spiral velocity. As we mentioned before, the torix is generally based on the helicola of infinite even dimensions. Thus, this torix would contain an infinite number of spirals wound one around another, with each preceding spiral serving as a leading spiral for the next trailing spiral. To maintain the integrity of the torix, each trailing spiral must propagate faster than its leading spiral. As the dimensional level of the torix approaches infinity, the spiral velocity of the last outer spiral approaches the ultimate spiral field velocity C that is only slightly greater than the velocity of light.

It is a monumental task to describe mathematically a torix of infinite even dimensions. Fortunately, we do not have to do so. A high accuracy of calculations can be achieved by considering the torix of the fourth dimension, and assuming that the spiral velocity of the trailing spiral V_2 is equal to the ultimate spiral field velocity C. Therefore, according to the Theorem of Pythagoras, the spiral velocity of the trailing spiral V_2 is related to its rotational and translational components by the equation:

$$V_2 = C = \sqrt{V_{2t}^2 + V_{2r}^2} \qquad \text{(D7)}$$

The Helix

Geometrically, the helix is based on the helicola of infinite odd dimensions. For the sake of simplicity, we showed in Fig. D5 a helix based on the helicola of the fifth dimension. This helix appears as double helical spiral[5]. Similarly to the double toroidal spiral, constructing a double helical spiral involves engaging a rod *ab* in two simultaneous motions, rotational and translational. The rotational motion of the rod *ab* is around its middle point with the rotational velocity V_{3r}. For an observer located at the middle point, the end points of the rod, *a* and *b*, will move along a circle with the radius r_3. During translational motion, the rod's middle point moves along a straight line with translational velocity V_{3t}.

As a result of the simultaneous rotational and translational motions of the rod *ab*, the end points of the rod will move along the two leading helical spirals A_3. The two leading helical spirals are angularly separated from one another by 180 degrees and form the leading double helical spiral. The helix comprises one winding of the leading double helical spiral.

Construction of a trailing double helical spiral involves engaging a rod *cd* in two simultaneous motions, rotational and translational. The rotational motion of the rod *cd* is around its middle point with the rotational velocity V_{4r}. For an observer located at the middle point, the end points of the rod *c* and *d* will move along a circle with radius r_4. During translational motion, the middle point of the rod *cd* moves along the leading helical spiral A_3 with translational velocity V_{4t}. As a result of the simultaneous rotational and translational motions of the rod cd, the end points of the rod will move along two trailing helical spirals A_4. The two trailing helical spirals are angularly separated one from another by 180 degrees and form the trailing double helical spiral.

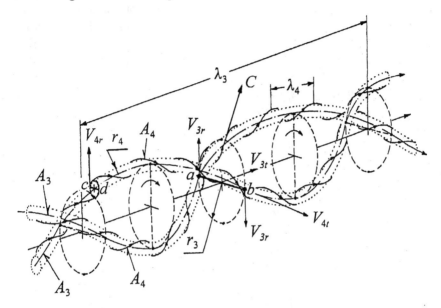

Fig. D5 Construction of a helix.

Similarly to the torix, there are two principal geometrical features of the helix that enable it to become one of the two prime elements of nature:

1. Length of one winding of the trailing spiral L_4 is equal to the length of the leading spiral L_3:

$$L_4 = L_3 = \frac{2\pi r_3^2}{r_4} \tag{D8}$$

2. For a group of helices that belong to the same family, the inversion radius r_i remains constant:

$$r_i = r_4 - r_3 = const. \tag{D9}$$

As was in the case with the torix, we do not have to describe mathematically the helicola of infinite odd dimensions. A high accuracy of calculations can be achieved by considering the helix of the fifth dimension and assuming that the spiral velocity of the trailing spiral V_4 is equal to the ultimate spiral field velocity C. Therefore, according to the Theorem of Pythagoras, the spiral velocities of the leading and trailing spirals are related to each other by the equations:

$$V_4 = C = \sqrt{V_{4t}^2 + V_{4r}^2} \tag{D10}$$

$$V_{4t} = \sqrt{V_{3t}^2 + V_{3r}^2} \tag{D11}$$

Translational velocity of the leading spiral V_{3t} is equal to the velocity of light c. The velocity of light c is only slightly smaller than the ultimate spiral field velocity C. For visible light in vacuum:

$$c = 0.999999999999997C.$$

REFERENCES

1. Born, M., *Atomic Physics*, Dover Publications, Inc., New York, 1969.
2. Domb, C., *Clerk Maxwell and Modern Science*, University of London, The Athlone Press, 1963.
3. Tricker, R. A. R., *The Contribution of Faraday and Maxwell to Electric Science*, Pergamon Press, Oxford, London, 1966.
4. Lewis, H., *Geometry - A Contemporary Course*, Third Edition, McCormick-Mathers Publishing Company, Third Edition, Cincinnati, Ohio, 1973.
5. Ginzburg, V.B., "Toroidal Spiral Field Theory," *Speculations in Science and Technology*, Vol. 19, 1996.

E

THE TORIX AND THE HELIX

The Appearance of the Torix

The appearance of the torix is convenient to describe as a function of the relative radius of its leading circle b_1 that is equal to the ratio of the radius of the leading circle r_1 to the inversion radius r_i:

$$b_1 = \frac{r_1}{r_i} \qquad \text{(E1)}$$

Chapter 2 of this book illustrates various appearances of the torus as its relative spiral radius b_1 changes from positive infinity to negative infinity. There are three major metamorphoses of the torus shape[1]:

   Change of direction of vorticity - This change is easy to visualize for real torices. For these torices, it occurs at $b_1 = 1$ when the torix reduces to a so-called *Zero-Point-Energy* (*ZPE*) *ring*. During this change, the left-handed torix becomes right-handed, and vice versa. The change of the direction of vorticity is accompanied by a change of the sign of the spiral radius of the trailing spiral r_2 and the sign of the rotational velocity of the trailing spiral V_{2r}.

   It is rather easy to visualize the change in the direction of vorticity of the torices because we are quite familiar with the left-handed and right-handed spirals. One can simply take a piece of thin wire and wind it around a rod in the form of either left-handed or right-handed spiral. In the SFT, there is a little bit more to it. To understand the difference, replace a thin wire with a narrow thin strip with its one side painted green and the other side red. Attach one end of the strip to the rod and wind the strip around the rod, so it will appear, for instance, as a green left-handed spiral. Unwrap the strip without detaching its fixed end and then wrap it around the rod forming the right-handed spiral. It will now look red. Thus, in this case, the spiral changes not only its spirality from left- to right-handed, it also becomes *inverted inside out*.

Change of reality - This change occurs at $b_1 = 0.5$. When $b_1 > 0.5$, all the parameters of the torices are expressed with real numbers. These torices are called *real torices*. When $b_1 < 0.5$, some parameters of the torices, including the translational velocity of the torix trailing spiral V_{2t}, are expressed with imaginary numbers. These torices are called the *imaginary torices*.

What is the physical meaning of the real and imaginary spiral fields? One interpretation would be that the real spiral field *thrusts into* the surrounding media, while surrounding media *is sucked into* the imaginary spiral field. This process may remind us of exhaling and inhaling air. Another interpretation would be that the real spiral field travels *forwards in time*, i.e. travels in a conventional way, while the imaginary spiral field travels *backwards in time*. Although the notion of the spiral field traveling backward in time sounds very strange, this concept is far from being novel to scientists.

Maxwell's famous equations yield two solutions, one equivalent to a positive energy wave flowing into the future, and the other describing a negative energy wave flowing into the past[2]. In a series of papers published between 1929 and 1932, the Dutch physicist Adriaan Fokker brought attention of the scientists to the fact that Maxwell's equations are completely symmetrical in time. In 1940, John Wheeler and Richard Feynman further expanded this concept in their absorber theory[2,3], which explains instantaneous transmission of light. The principal assumption of this theory is that an excited electron emits two types of waves, the "retarded" waves that travel forwards in time, and the "advanced" waves that travel backwards in time.

Change of sign of radius of the torix leading spiral - This change occurs only in the imaginary torices at $b_1 = 0$ when the torix reduces to a so-called *Maximum-Point-Energy (MPE) ring*. We are familiar with the notion of a negative number. If, for example, a sea level represents zero level then the height of a hill will be expressed by positive numbers, while the depth of the valleys by negative numbers. Let us extend this interpretation of the positive and negative numbers to two other examples that are closer to the SFT. In the first example, use a thin flat elastic membrane with one side painted green and the other one red. Take a plate with a round hole, and attach the membrane to the plate so it will cover the hole, appearing green if you look from the top. If you will inflate the membrane from the bottom then it will appear as a green semi-balloon with positive radius. Conversely, if you will inflate the membrane from the top then it will appear as a red semi-balloon with negative radius. Thus, in this example, changing the sign of the radius of the semi-balloon would indicate that the semi-balloon was *inverted inside out*.

In the second example, we will use a complete thin plastic balloon, painted green outside and red inside. After inflating the balloon, it will appear green with the positive radius. After deflating the balloon, its radius will become theoretically equal to zero and its color will disappear. What shall we do to make this balloon red, and, thus to have a negative radius? The answer is rather simple. Invert it inside out and inflate again. Thus, the torus with negative spiral radius is the torus that is *inverted inside out*.

Wave and Particle Properties of the Torix and the Helix

Both the torix and the helix have certain properties that can identify them as either the waves or particles. As waves, they have wavelength and frequency. As particles, they have gravitational mass, inertial mass, angular momentum, spin, and electric charge. Other properties, such as velocity and energy of the torix and helix, belong to both waves and particles. At the same time, the helix parameters are closely related to the parameters of the torix from which they are originated.

Table E1 Relativistic equations of the SFT for real and imaginary torices.

Torix parameter	Equation	Comments for torices	
		Real	Imaginary
Relative gravitational mass	$\dfrac{m_{tg}}{m_o} = abs\,\dfrac{\sqrt{1 - (V_{2t}/C)^2}}{2}$	Gravitational mass decreases with velocity	Gravitational mass increases with velocity
Relative electric charge	$\dfrac{e_t}{e_o} = \pm\,\dfrac{\sqrt{1 - (V_{2t}/C)^2}}{2}$	Electric charge decreases with velocity	Electric charge increases with velocity

Similarly to the theory of relativity[4], the SFT establishes the relativistic relationship between the parameters of both the torix and helix. As shown in Table E1, the relativistic equations of the SFT are quite different than those used in the theory of relativity. According to the SFT, both the relative gravitational mass m_{tg}/m_o and the relative electric charge e_t/e_o of the real torices decrease with the increase of its translational velocity V_{2t}, while in the imaginary torices the same parameters increase with the increase of its translational velocity V_{2t}. This is due to the fact that the square of the translational velocity V_{2t} is positive for the real torices while it is negative for the imaginary torices.

In Section D, we presented the spiral field as a trace left by the ends of the rod that is involved in two motions, rotational and translational, as shown in Figs. D3 and D4. After establishing the relativistic relationships of the torix and the helix, one can give another interpretation of the physical meaning of the spiral field. One may think that the spiral field can be represented by the traces that are left by the electric charges of an electric dipole that is involved in both the rotational and translational motions.

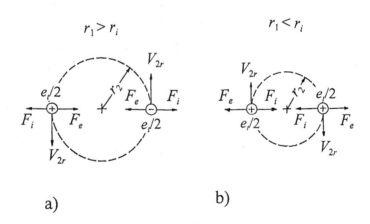

a)   b)

Fig. E1 Representation of the spiral field as a trace left by the electric charges of a rotating electric dipole.

As an example, let us consider two electric dipoles representing the torix trailing spirals with the spiral radius r_2. The electric dipoles are involved in rotational motion only with the rotational velocity V_{2r} as will be seen by an observer located at the middle of the electric dipoles. The spiral radius of the leading circle r_1 of the first dipole (Fig. E1a) is greater than the inversion radius r_i. This corresponds to the left-handed toroidal spiral that can be envisioned as a trace left by the opposite electric charges of the electric dipole. The electric charges $e_2/2$ will stay in equilibrium with one another due to the balance between the electric (centripetal) force F_e and the inertial (centrifugal) force F_i.

The spiral radius of the leading circle r_1 of the second dipole (Fig. E1b) is less than the inversion radius r_i. This corresponds to the right-handed toroidal spiral that can be envisioned as a trace left by the electric charges of the electric dipole that have the same sign. The electric charges $e_2/2$ will stay in equilibrium with one another due to the balance between the electric (centrifugal) force F_e and the inertial (centripetal) force F_i. Since the electric charges of the electric dipole propagate along a spiral path with the ultimate spiral field velocity C, their magnitudes along the spiral path will be equal to zero in compliance with the relativistic equations shown in Table E1.

The effect of geometry on the property of the torix and the helix is convenient to express as a function of the *vortex ratio* that is a ratio of the spiral radii of the trailing and leading spirals. In application to the torix and helix, the vortex ratios x_t and x_h are respectively equal to:

$$x_t = \frac{r_2}{r_1} \qquad x_h = \frac{r_4}{r_3} \tag{E2}$$

Then the relative gravitational masses and electric charges of the torix and the helix can be represented by the equations shown in Table E2.

Table E2 Torix and helix gravitational masses and electric charges.

Parameter	Torix	Helix
Relative gravitational mass	$\dfrac{m_{tg}}{m_o} = abs\left(\dfrac{x_t}{2}\right)$	$\dfrac{m_{gh}}{m_o} = abs\left(\dfrac{x_h}{4}\sqrt{1 - x_h^4}\right)$
Relative electric charge	$\dfrac{e_t}{e_o} = \pm abs\left(\dfrac{x_t}{2}\right)$	$\dfrac{e_h}{e_o} = \pm abs\left(\dfrac{x_h}{4}\sqrt{1 - x_h^4}\right)$

Since the vortex ratios x_t and x_h represent the curvature of the torices and helices, one can think that both the gravitational masses and electric charges are the functions of the curvature of the spiral field. This may remind us of the theory of relativity, which considers the gravity to be the result of the curvature of empty space[4]. The other important parameters of both torices and helices are inertial mass, angular momentum, and spiral frequency, which are described in Chapters 4 and 5 of this book. Here, we will identify some parameters of the torix and the helix that are needed for our discussion.

Torix inversion radius r_i is equal to:

$$r_i = \frac{ke_o^2}{2m_oC^2} \tag{E3}$$

where

m_o, e_o = rest mass and rest electric charge of a particle made of one real and one imaginary torices

k = Coulomb's constant.

Fundamental spiral field frequency f_o is given by:

$$f_o = \frac{m_oC^3}{k\pi e_o^2} \tag{E4}$$

Torix relative spiral frequency δ_2 is equal to:

$$\delta_2 = f_2/f_o = \frac{1}{b_1} = \frac{r_i}{r_1} \tag{E5}$$

Notice that the fundamental spiral frequency is equal to the torix spiral frequency when the radius of its leading circle r_1 is equal to the inversion radius r_i.

Quantum Energy Levels of the Torix

Two kinds of processes produce quantum changes in the physical properties of the torix[1]. These processes are described in Chapter 6 of this book, where they are defined as *oscillation* and *excitation processes*.

The oscillation process - During the oscillation process, the torix inversion radius r_i changes, while the other geometrical parameters remain relatively the same in respect to the inversion radius. Subsequently, as a result of the oscillation process, all the geometrical parameters of the torix change proportionally by the same scale factor that is called the *torix oscillation factor Q*. The first three oscillation levels $N = 0$, 1, and 2 are called *harmonic*, while the higher oscillation levels are called *universal*.

The oscillation process produces a pronounced effect on the torix physical properties. For instance, as shown in Table E3, when the torix inversion radius r_{iN} increases by the torix oscillation factor Q, the torix rest mass m_{oN}, as well as the fundamental spiral field frequency f_{oN} decrease by the same factor. At the oscillation energy level $N = 27$ corresponding to the maximum value of the torix oscillation factor Q, the torix has the largest rest mass of 1.356×10^{11} MeV/c^2, the highest fundamental spiral frequency of 8.984×10^{33} Hz, and the smallest inversion radius $r_i = 5.311 \times 10^{-27}$ m.

Table E3 Parameters of oscillation torices ($0 \leq N \leq 27$).

Level type	N	Q	m_{oN} MeV/c^2	f_{oN} Hz	r_{iN} m
Harmonic	0	1.000	0.511	3.386×10^{22}	1.409×10^{-15}
	1	2.000	1.022	6.772×10^{22}	7.045×10^{-16}
	2	3.000	1.533	1.016×10^{23}	4.697×10^{-16}
Universal	3	205.554	105.038	6.961×10^{24}	6.855×10^{-18}
	4	3521.04	1799.246	1.192×10^{26}	4.002×10^{-19}
	5	35741.4	18263.815	1.210×10^{27}	3.942×10^{-20}
	..	....	....	....	....
	27	2.653×10^{11}	1.356×10^{11}	8.984×10^{33}	5.311×10^{-27}

The excitation process - During the excitation process, the torix spiral radius r_1 changes, while its inversion radius r_i remains constant. There are two types of the excitation process, universal and harmonic. The torices involved in these processes are respectively called *harmonic torices* and *universal torices*. The properties of universal torices are expressed as a function of the *universal quantum parameter z*. This parameter is the product of the *torix universal excitation level n* and the *universal spiral field constant U*. It is given by the equation[1]:

$$z = nU \qquad n = 0, 1, 2, \ldots \qquad \text{(E5)}$$

Notice that the inverse of the universal spiral field constant U is the fine structure constant α^5.

As an example, Table E4 shows the relative spiral radii b_1 of the real and imaginary universal torices.

Table E4 Relative spiral radii of some universal torices.

	Real torices			Imaginary torices	
	Negative	Positve		Negative	Positive
n	$2 \leq b_1 < \infty$	$1/2 \leq b_1 \leq 2/3$	n	$-\infty < b_1 \leq -2$	$2/5 \leq b_1 \leq 1/2$
0	2.000	2/3	0	- 2.000	0.40000000
1	37559.72	0.50000666	1	- 37559.72	0.49999334
2	150232.90	0.50000166	2	- 150232.90	0.49999834
3	338021.52	0.50000074	3	- 338021.52	0.49999926
4	600925.58	0.50000042	4	- 338025.58	0.49999958

Properties of the harmonic torices are expressed as a function of the *torix harmonic excitation levels m* = 0, 1, 2... As an example, Table E5 shows the relative spiral radii b_1 of the real and imaginary harmonic torices.

The oscillation and excitation processes create sufficient variety of torices to assemble all the known particles, as will be described later.

Table E5 Relative spiral radii of some harmonic torices.

	Real torices			Imaginary torices	
	Negative	Positive		Negative	Positive
m	$2 \leq b_1 < \infty$	$1/2 \leq b_1 \leq 2/3$	m	$-\infty < b_1 \leq -2$	$2/5 \leq b_1 \leq 1/2$
0	2	2/3	0	-2	2/5
1	3	3/5	1	-3	3/7
2	4	4/7	2	-4	4/9
3	5	5/9	3	-5	5/11
4	6	6/11	4	-6	6/13

The Creation of Torices

Torices were originally created from empty space by three kinds of polarization processes:

- Polarization of the first kind produces pairs of *reality-polarized torices* with mutually reverse flow of energy, outward and inward. The torices with outward flow of energy are called the *real torices*, while the torices with inward flow of energy are called the *imaginary torices*. In real torices, time flows forwards, while in imaginary torices, time flows backwards.
- Polarization of the second kind divides both the real and imaginary torices into pairs of *vorticity-polarized torices* with mutually inverse vorticities of their first trailing spirals, right-handed and left-handed. Respectively, the electric charges of these torices are considered positive and negative
- Polarization of the third kind further splits the torices with the same direction of vorticity into pairs of *complementary-polarized torices*, outer and inner. The outer torices form the particles of matter while the inner torices form the particles of the dynamic ether.

Polarization of the first kind begins from a creation of a *primordial spherix* containing both real and imaginary torices at their respective extreme levels, ZPE ring and MPE ring, as shown in Fig. E2. The latter can be thought as a *micro black hole*. Both rings correspond to the highest torix oscillation level N, and, therefore, have maximum masses and minimum inversion radii. It shall be noted that because of the quantum character of the spiral field, the torices never reduce to the ZPE and MPE rings, but asymptotically approach to these states.

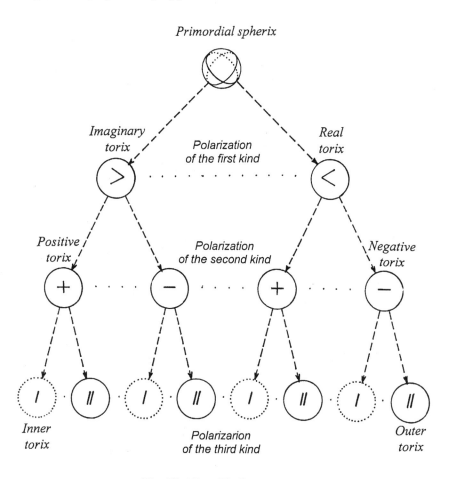

Fig. E2 Micro-Big-Bang process.

The process of the oscillation of the torix from the highest oscillation level to a lower oscillation level is called the *Micro-Big-Bang inflation process*, echoing the cosmological Big-Bang inflation process proposed by Guth[3]. During the Micro-Big-Bang inflation process, a micro black hole maintains its communication with the *point of origin*. The communication is practically instantaneous due to the fact that the translational velocity of the MPE ring is approaching infinity. This feature assures practically instantaneous communication between all the micro black holes that were originated from the same point of origin, with the point of origin serving as a communication center.

The Formation of Helices

When a torix is transferred from a higher to lower energy level, there will be emitted two principal types of helices, real (*retarded*) and imaginary (*advanced*). The real helices are associated with real torices and propagate forward in time, while the imaginary helices are associated with imaginary torices. and propagate backwards in time. The real and imaginary helices are functionally similar to the retarded and advanced waves of the absorber theory proposed by Wheeler and Feynman[2,3].

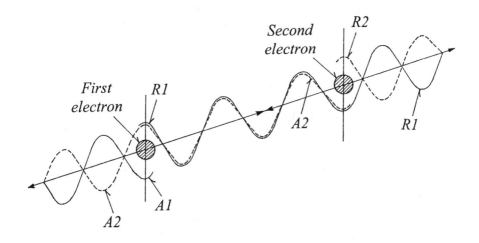

Fig. E3 Instantaneous transmission of light.

Consider two electrons separated by a certain distance (Fig. E3). After its excitation, the first electron emits simultaneously a half retarded wave *R1* and a half advanced wave *A1* that propagate in opposite directions with the velocity of light. After the half retarded wave *R1* reaches the second electron, this electron will emit in response a half retarded wave *R2* and a half advanced wave *A2* that propagate in opposite directions with the velocity of light. The two sets of waves cancel out everywhere except in the region between the two electrons. Since within this region the half retarded wave *R1* travels forward in time and the half advanced wave *A2* travels backwards in time, the connection between the two electrons is made simultaneously.

Both the real and imaginary helices are divided into two groups, fast and slow. The translational velocity of fast helices is equal to the velocity of light *c* that is only slightly slower than the ultimate spiral field velocity *C*. The slow helices propagate at a much slower pace. Remarkably, in spite of the substantial difference in their translational velocities, both fast and slow helices have the same spiral frequency.

Both fast and slow helices are divided into two subgroups, oscillation and excitation. Oscillation helices are either emitted or absorbed during oscillation of torices, i.e. when the torices change their oscillation level *N*. Excitation helices are either emitted or absorbed when the torices change their excitation levels, either harmonic *m* or universal *n*. Notably, under comparable conditions, the spiral frequencies of the oscillation helices are generally greater than those of the excitation helices.

REFERENCES

1. Ginzburg, V.B. "Dynamic Aether,"*Journal of New Energy*, Vol. 6, No. 1, 2001.
2. Gribbin, J., *Q Is For Quantum - An Encyclopedia of Particle Physics*, A Touchstone Book, Simon & Schuster, New York, 2000.
3. Price, H., *Time's Arrow and Archimedes' Point - New Directions for the Physics of Time*, Oxford University Press, New York, 1966.
4. Einstein, A., *Relativity – The Special and the General Theory*, Crown Publishers, Inc., 1952.
5. Born, M., *Atomic Physics*, Dover Publications, Inc., New York, 1969, p. 169.

F

ELEMENTARY PARTICLES

The Structure of Elementary Particles

One can consider the torices located at various oscillation and excitation levels as the building blocks from which the elementary particle are formed. Obviously, the fact that the elementary particles are made of torices makes the adjective "elementary" obsolete in application to the elementary particles. Designation of torices is shown below.

A, a, B, b, etc. . . . . Outer complementary-polarized torices

A, a, B, b, etc. . . . . .Inner complementary-polarized torices

A, A, B, B . . . . .Real torices

$\overline{A}, \overline{A}, \overline{B}, \overline{B}, \overline{a}, \overline{a}, \overline{b}, \overline{b}$, etc. . . . .Imaginary torices.

The first subscript indicate the torix oscillation level N, and the second subscript indicates the excitation level, either universal n or harmonic m.

Examples:

a) $+A_{02}$ is the real positive outer torix at the oscillation level $N = 0$, and excitation level $n = 2$, or $m = 2$;

b) $-\overline{B}_{31}$ is the imaginary negative inner torix at the oscillation level $N = 3$, and excitation level $n = 1$, or $m = 1$.

The elementary particles are made of torices located at various oscillation levels N and at various excitation levels either harmonic m or universal n. In general, elementary particles may contain both a core and a shell. The core is made of the torices located at the higher oscillation level N. It is electrically neutral and contains the greater portion of the particle mass. The shell is made of the torices located at the lower oscillation level N, and its electric charge can alternatively be positive, negative, or neutral. Only the particles that are made of both the real and imaginary torices can be stable. The most stable are the particles in which the total relative spiral energy of their torices ϵ_t is equal to zero.

Several examples of the most simple elementary particles are shown in the following sections. There, the calculated values of three principal parameters of the particles are given in comparison with the available experimental values:

- Gravitational mass m_g
- Relative electric charge e/e_o
- Magnetic moment ratio μ/μ_b.

Particle magnetic moment μ is determined as an algebraic sum of the magnetic moments of the real torices that make up the particle shell. For the real torix with spiral radius r_1 and electric charge e_t moving with translational velocity V_{2t}, the magnetic moment μ_t is equal to:

$$\mu_t = e_t V_{2t} (r_1 - r_i) \tag{F1}$$

Conventionally, the particle magnetic moments μ are expressed as a ratio μ/μ_b where μ_b is the base magneton that is given by[1-3]:

$$\mu_b = \frac{e_o h}{4\pi m_{ob}} \tag{F2}$$

where

e_o = electron rest electric charge
h = Planck constant.

For electrons and positrons. . . .$m_{ob} = m_{oe}$ = rest mass of electron
For muons . . . . $m_{ob} = m_{o\mu}$ = rest mass of muon
For protons and neutrons . . . . .$m_{ob} = m_{op}$ = rest mass of proton
For tau particles. . . . . $m_{ob} = m_{o\tau}$ = rest mass of tau.

When $m_{ob} = m_{oe}$, the base magneton μ_b is known as the *Bohr magneton* μ_B. When $m_{ob} = m_{op}$, the base magneton μ_b is known as the *nuclear magneton* μ_N.

Leptons

According to the SFT, leptons are coreless. They are made of torices located at the universal excitation level $n = 1$ that is also known as the *ground level*, as shown in Tables F1-F8 and Fig. F1. The torices of electrons and positrons are located at the lowest oscillation level $N = 0$, and the torices of heavier leptons are located at higher oscillation levels N.

Table F1 Electron e^- and positron e^+ ($N = 0$, $n = 1$, $m_{oe} = 0.5109907$ MeV/c^2).

Torix	b_1	No. tor.	μ/μ_b	m_g MeV/c^2	e/e_o
$\pm A_{01}$	37,559.724	1	± 0.9999667	0.2554885	± 0.4999867
$\pm \bar{A}_{01}$	-37,559.724	1	-	0.2555022	± 0.5000133
Calculated values:			± 0.9999667	0.51099907	± 1.000
Measured values[1]:			± 1.0011597	0.51099907	± 1.000

Table F2 $3e^-$ and $3e^+$ ($N = 2$, $n = 1$, $m_{o3e} = 1.5299721$ MeV/c^2).

Torix	b_1	No. tor.	μ/μ_b	m_g MeV/c^2	e/e_o
$\pm A_{21}$	37,559.724	1	± 0.9999667	0.7664656	± 0.4999867
$\pm \bar{A}_{21}$	-37,559.724	1	-	0.7665065	± 0.5000133
Calculated values:			± 0.9999667	1.5329721	± 1.000

Table F3 Muon μ^- and muon μ^+ ($N = 3$, $n = 1$, $m_{o\mu} = 105.038$ MeV/c^2).

Torix	b_1	No. tor.	μ/μ_b	m_g MeV/c^2	e/e_o
$\pm A_{31}$	37,559.724	1	± 0.9999667	52.5176	± 0.4999867
$\pm \bar{A}_{31}$	-37,559.724	1	-	52.5204	± 0.5000133
Calculated values:			± 0.9999667	105.038	± 1.000
Measured values[1]:			± 1.0011659	105.658389	± 1.000

Table F4 Tau τ^- and tau τ^+ ($N = 4$, $n = 1$, $m_{\varrho\tau}$ =1799.246 MeV/c²).

Torix	b_1	No. tor.	μ/μ_b	m_g MeV/c²	e/e_o
$\pm A_{41}$	37,559.724	1	$\pm$ 0.9999667	899.59905	$\pm$ 0.4999867
$\pm \overline{A}_{41}$	-37,559.724	1	-	899.64695	$\pm$ 0.5000133
			$\pm$ 0.9999667	1799.246	$\pm$ 1.000
Measured values[1]:			-	1777.00	$\pm$ 1.000

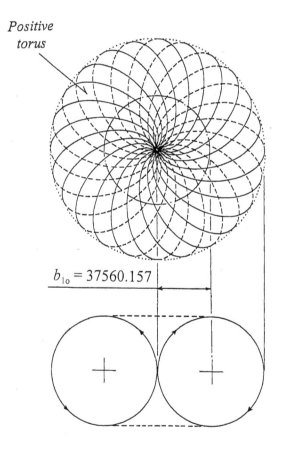

Positive torus

$b_{1o} = 37560.157$

Fig. F1 Real tori of a lepton at the exitation level $n = 1$.

Etherons

According to the SFT, etherons are made of inner complementary torices located at the torix universal excitation levels $n \geq 1$. They are coreless, as shown in Tables F5-F8 and Fig. F2.

Table F5 e-based etherons ($N = 0$, $n = 1$, $m_o = 0.5109907$ MeV/c^2).

Torix	b_1	No. tor.	μ/μ_b	m_g MeV/c^2	e/e_o
$\pm A_{01}$	1.00002663	1	$\pm 2.59 \times 10^{-12}$	0.0000068	± 0.00001331
$\pm \bar{A}_{01}$	-1.00002663	1	-	0.5109839	± 0.99998669
	Calculated values:		$\pm 2.59 \times 10^{-12}$	0.5109907	± 1.000

Table F6 $3e$-based etherons ($N = 2$, $n = 1$, $m_{o3e} = 1.5299721$ MeV/c^2).

Torix	b_1	No. tor.	μ/μ_b	m_g MeV/c^2	e/e_o
$\pm A_{21}$	1.00002663	1	$\pm 2.59 \times 10^{-12}$	0.0000205	± 0.00001331
$\pm \bar{A}_{21}$	-1.00002663	1	-	1.5299516	± 0.99998669
	Calculated values:		$\pm 2.59 \times 10^{-12}$	1.5299721	± 1.000

Table F7 Muon-based etherons ($N = 3$, $n = 1$, $m_{o\mu} = 105.038$ MeV/c^2).

Torix	b_1	No. tor.	μ/μ_b	m_g MeV/c^2	e/e_o
$\pm A_{31}$	1.00002663	1	$\pm 2.59 \times 10^{-12}$	0.001408	± 0.00001332
$\pm \bar{A}_{31}$	-1.00002663	1	-	105.03659	± 0.99998669
	Calculated values:		$\pm 2.59 \times 10^{-12}$	105.038	± 1.000

The etherons are the prime elementary particles. At the onset of the Micro-Big-Bang process, they form the primordial spherix in which they are located at the highest oscillation and excitation levels.

Table F8 Tau-based etherons ($N = 4$, $n = 1$, $m_{o\tau}$ =1799.246 MeV/c^2).

Torix	b_1	No. tor.	μ/μ_b	m_g MeV/c^2	e/e_o
$\pm A_{41}$	1.00002663	1	$\pm 2.59 \times 10^{-12}$	0.0241099	± 0.00001332
$\pm \overline{A}_{41}$	-1.00002663	1	-	1799.2219	± 0.99998669
	Calculated values:		$\pm 2.59 \times 10^{-12}$	1799.246	± 1.000

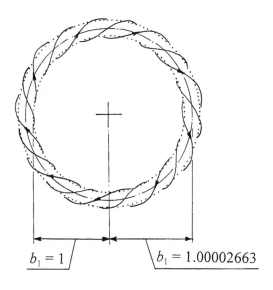

$$b_1 = 1 \qquad\qquad b_1 = 1.00002663$$

Fig. F2 Real tori of an etheron at the excitation level $n = 1$.

Nucleons

Proton and neutron are jointly known as *nucleons*. They have identical cores that are electrically neutral. Each core contains the largest portion of the nucleon's gravitational mass. As shown in Table F9 and Fig. F3, the real part of the nucleon core is made of three sets of real torices $+a_{30}$ and $-a_{30}$ that are angularly spaced in 120 degrees. Positive and negative torices are angularly spaced next to each other. They are located in two mutually perpendicular planes X and Y, with two positive torices and one negative torix in one plane, and two negative torices and one positive torix in the other plane. The imaginary part of the nucleon core is made of three sets of imaginary torices $+\bar{a}_{30}$ and $-\bar{a}_{30}$. All the torices of the nucleon core are located at the oscillation level $N = 3$ and the excitation level $n = 0$.

Table F9 Nucleon core ($N = 3$, $n = 0$, $m_o = 105.038$ MeV/c^2).

Torix	b_1	b_{1o}	No. toric.	ϵ_t	m_{tg} MeV/c^2	e/e_o
$-a_{30}$	2/3	0.0032432	3	4.5	78.7785	-0.75
$+a_{30}$	2/3	0.0032432	3	4.5	78.7785	+0.75
$-\bar{a}_{30}$	-2/3	-0.0032432	3	-4.5	393.8925	-3.75
$+\bar{a}_{30}$	-2/3	-0.0032432	3	-4.5	393.8925	+3.75
		Calculated values:		0.00	945.342	0.00

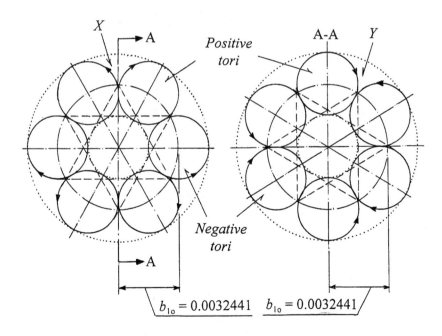

Fig. F3 Arrangement of the real tori of nucleon core.

It is important to remember that the absolute values of the torix spiral radius r_1 depend on the torix oscillation levels N. This means that to calculate the absolute value of the spiral radius r_1, one must multiply the relative spiral radius b_1 by the inversion radius r_{iN} corresponding to the oscillation level N. To compare the dimensions of the torices that belong to different oscillation levels N, it is convenient to express the relative spiral radii of the torices in the same units as the relative spiral radii b_{1o} that corresponds to the oscillation level $N = 0$.

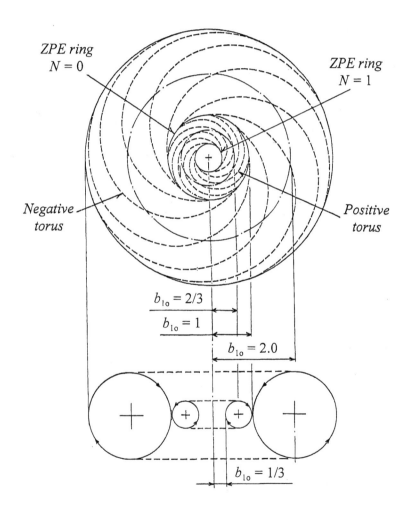

Fig. F4 Real tori of the neutron shell.

For instance, in Table F9, the torix relative spiral radius b_1 is expressed in respect to the inversion radius r_i of the torix located at the oscillation level $N = 3$. To calculate the relative spiral radii b_{1o}, one shall divide b_1 by the torix oscillation factor $Q = 205.554$ (see Table E3).

The Neutron

The neutron is made of a core and a shell. The structure of the core is shown in Table F9 and Fig. F3. The shell is electrically neutral as shown in Table F10 and Fig. F4. It is made of two sets of torices, negative and positive. The negative set comprises one real and one imaginary torices $-A_{00}$ and $-\bar{A}_{00}$ with both torices located at the oscillation level $N = 0$ and the excitation level $n = 0$. The positive set comprises one real and one imaginary torices $+A_{20}$ and $+\bar{A}_{20}$ with both torices located at the oscillation level $N = 2$ and the excitation level $n = 0$.

Table F10 Neutron shell $(n = 0)$.

Torix	N	m_o MeV/c^2	b_{1o}	No. tor.	μ/μ_b	m_{tg} MeV/c^2	e/e_o
$-A_{00}$	0	0.51099907	2.0	1	-2.9009637	0.127750	-0.25
$-\bar{A}_{00}$	0	0.51099907	-2.0	1	-	0.383249	-0.75
$+A_{20}$	2	1.5299721	2/3	1	+0.9669879	0.383249	+0.25
$+\bar{A}_{20}$	2	1.5299721	-2/3	1	-	1.149748	+0.75
				Calculated values:	-1.9339758	2.043996	0.00
				Measured values[1]:	-1.9130428	-	0.00

From Tables F9 and F10, we find that the total calculated gravitational mass of the neutron is equal to: $945.342 + 2.044 = 947.386$ MeV/c^2. Compare with the measured neutron mass[1] of 939.56563 MeV/c^2.

The Proton

The proton is made of a core and a shell. The structure of the proton core is identical to that of the neutron, and shown in Table F9 and Fig. F3. The proton shell is formed as a result of the decay of a neutron shell. Two kinds of transformations of the torices take place during the decay:

1. The oscillation level of positive torices $+A_{20}$ and $+\bar{A}_{20}$ reduces from $N = 2$ to $N = 0$. Subsequently, they transform respectively into positive torices $+A_{00}$ and $+\bar{A}_{00}$, forming the proton shell (see Table F11 and Fig. F5).

2. The universal excitation level of negative torices $-A_{00}$ and $-\bar{A}_{00}$ increases from $n = 0$ to $n \to \infty$. Subsequently, they transform respectively into negative torices $-A_{0\infty}$ and $-\bar{A}_{0\infty}$, forming a free electron (see Table F1).

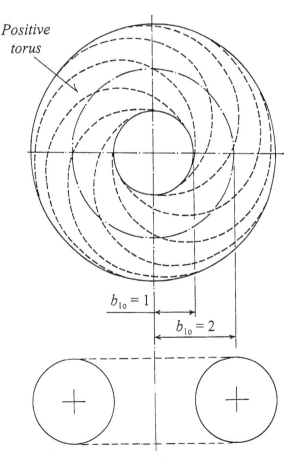

Fig. F5 Real torus of nuclear positron.

From Tables F9 and F11, we find that the total calculated gravitational mass of the proton is equal to: 945.342 + 0.511 = 945.853 MeV/c². Compare with the measured proton mass[1] of 938.27231 MeV/c².

Table F11 Proton shell ($N = 0$, $n = 0$, $m_o = 0.51099907$ MeV/c²).

Torix	b_1	b_{1o}	No. tor.	ϵ_t	μ/μ_b	m_{tg} MeV/c²	e/e_o
$+A_{00}$	2.0	2.0	1	+0.5	+2.9009637	0.127750	+0.25
$+\overline{A}_{00}$	-2.0	-2.0	1	-0.5	-	0.383249	+0.75
Calculated values:				0.00	+2.9009637	0.51099907	+1.00
Measured values[1]:				-	+2.7928474	-	+1.00

Comparison With Measured Data

Table F12 shows a summary of the calculated and measured parameters of the leptons and the nucleons.

Table F12 Calculated and measured parameters of leptons and nucleons.

Particle	Gravitational mass MeV/c²		Magnetic moment ratio μ/μ_b	
	Calculated	Measured	Calculated	Measured
Electron	0.5109907	0.5109907	- 0.9999667	- 1.0011597
Muon	105.038	105.658389	- 0.9999667	- 1.0011659
Tau	1799.246	1777.00	- 0.9999667	-
Proton	945.853	938.27231	+2.9009637	+2.7928474
Neutron	947.386	939.56563	-1.9339758	-1.9130428

Differences between the calculated and measured gravitational masses of the particles shown in Table F12 can be attributed to the binding energy that holds their torices together. For instance, the binding energy of the proton torices is equal to: 945.853 - 938.272 = 7.581 MeV/c², while the binding energy of the neutron torices is equal to: 947.134 - 939.566 = 7.568 MeV/c². Differences between the calculated and measured magnetic moment ratios of the particles shown in Table F12 are probably due to neglecting a contribution of the magnetic spins to the total calculated magnetic moments.

Table F13 Calculated and measured parameters of nucleons.

Difference between gravitational masses of neutron and proton MeV/c²		Ratio of magnetic moments of proton to neutron μ_p/μ_n	
Calculated	Measured	Calculated	Measured
1.533	1.293	- 1.50	- 1.46

Particles and Antiparticles

SFT clearly defines the relationship between particles and antiparticles. Structurally, both the particles and antiparticles are identical. They are made of the same types of torices located at the same oscillation and excitation levels. The only difference is in the vorticities of their torices. The vorticities of the torices in antiparticles are opposite to the vorticities of the corresponding torices in particles.

This definition of the particles and antiparticles allows one to visualize not only the structure of the charged particles and antiparticles, like electron and positron, but also the neutral particles, like neutron and antineutron.

REFERENCES

1. *Physical Review D: Particles and Fields*, Volume 54, Third Series, Part 1, Review of Particle Physics, The American Physical Society, 1 July, 1996.
2. Serway, R.A., *Physics For Scientists & Engineers with Modern Physics*, Third Edition, Saunders College Publishing, Philadelphia, 1992.
3. Livesey, D.L., *Atomic and Nuclear Physics*, Blaisdell Publishing Company, Toronto, 1966.

G

PHOTONS AND NEUTRINOS

Definition of Photons and Neutrinos

Both photons and neutrinos are made of helices as shown in Table G1.

Photons - The helices that form photons originate from parent torices located at the lowest oscillation level $N = 0$. There are two types of photons, harmonic and universal. Harmonic photons are emitted when the harmonic excitation levels m of the parent torices reduce. Universal photons are emitted when the universal excitation levels n of the parent torices reduce.

Neutrinos - The helices that form neutrinos originate from parent torices located at various oscillation levels N. There are two types of neutrinos, harmonic and universal. Harmonic neutrinos are emitted when the harmonic oscillation levels $N \leq 2$ of the parent torices reduce. Universal neutrinos are emitted when the universal oscillation levels $N > 2$ of the parent torices reduce.

Table G1 Classification of photons and neutrinos.

Particle	Quantum levels of torices			
	$N \leq 2$	$N > 2$	m	n
Photon	-	-	Harmonic excitation	Universal excitation
Neutrino	Harmonic oscillation	Universal oscillation	-	-

Photons and neutrinos are made of either real (retarded) or imaginary (advanced) helices. They are called either *real* (*retarded*) photons and neutrinos or *imaginary* (*advanced*) photons and neutrinos. Real photons and neutrinos travel forward in time, while imaginary photons and neutrinos travel backwards in time.

As we discussed in Section E, there are two types of helices, fast and slow. When originated during the same change of energy levels of parent torices, both fast and slow helices have the same spiral frequency f_3. However, whereas the translational velocity of fast helices is very close to the ultimate spiral field velocity C, slow helices propagate at a much slower pace. Fast helices form fast photons and fast neutrinos. Conversely, slow helices form slow photons and slow neutrinos. Fast and slow helices are emitted in opposite directions, as shown in Fig. G1.

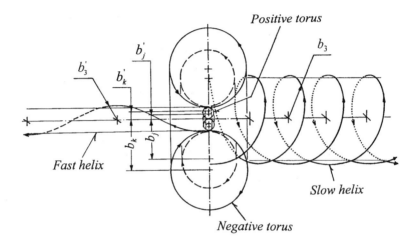

Fig. G1 Formation of fast and slow helices.

Universal Photons

Universal photons are emitted when parent universal torices shift from higher universal excitation levels n_k to lower levels n_j. If b_{1k} and b_{1j} are the spiral radii of the respective parent torices, then the spiral frequency of the emitted photon f_3 is equal to:

$$f_3 = \frac{m_o C^2}{h} \left(\frac{1}{b_{1j}} - \frac{1}{b_{1k}} \right) \qquad (G1)$$

Table G2 shows the parameters of the outer complementary-polarized torices $-A_{Nn}$ and $-\overline{A}_{Nn}$ of an atomic electron and the parameters of the inner complementary-polarized torices $-A_{Nn}$ and $-\overline{A}_{Nn}$ of a negative etheron at various universal excitation levels n. Tables G3 and G4 show the parameters of fast and slow photons that are emitted as a result of the shift of either the outer complementary-polarized torices $-A_{Nn}$ and $-\bar{A}_{Nn}$ or the inner complementary-polarized torices $-A_{Nn}$ and $-\bar{A}_{Nn}$ from higher universal excitation levels n_k to lower levels n_j.

Table G2 Relative spiral radii b_1 of universal torices of an atomic electron and a negative etheron ($N = 0$, $m_{oe} = 0.5109907$ MeV/c²).

Universal excitation level n	Atomic electron		Negative etheron	
	$-A_{Nn}$	$-\bar{A}_{Nn}$	$-A_{Nn}$	$-\bar{A}_{Nn}$
0	2.0	-2.0	2.0	-2.0
1	375559.72	-75559.72	1.00002663	-1.00002663
2	150232.90	-150232.90	1.00000666	-1.00000666
3	338021.52	-338021.52	1.00000296	-1.00000296
4	600925.58	-600925.58	1.00000166	-1.00000166

Table G3 Parameters of fast photons associated with the outer complementary-polarized torices $-A_{Nn}$ and $-\bar{A}_{Nn}$ of an atomic electron and the inner complementary-polarized torices $-A_{Nn}$ and $-\bar{A}_{Nn}$ of a negative etheron ($N = 0$, $m_{oe} = 0.5109907$ MeV/c²).

Universal excitation levels		Fast photon parameters $(V_{3t} \sim C)$			
n_k	n_j	b_3	f_3, Hz	m_{hg}, MeV/c²	Photon type
1	0	1.0009137400	3.090×10^{19}	2.332×10^{-04}	gamma-rays
2	1	1.0000000729	2.467×10^{15}	1.861×10^{-08}	visible light
3	2	1.0000000135	4.569×10^{14}	3.447×10^{-09}	infrared light
4	3	1.0000000047	1.599×10^{14}	1.207×10^{-10}	infrared light

Table G4 Parameters of slow photons associated with outer complementary-polarized torices $-A_{Nn}$ and $-\bar{A}_{Nn}$ of an atomic electron and inner complementary-polarized torices $-A_{Nn}$ and $-\bar{A}_{Nn}$ of a negative etheron ($N = 0$, $m_{oe} = 0.5109907$ MeV/c²).

Universal excitation levels		Slow photon parameters $(V_{3t} \ll C)$			
n_k	n_j	b_3	f_3, Hz	V_{3t}, m/s	m_{hg}, MeV/c²
1	0	167.6125	3.089×10^{19}	5.032×10^{08}	3.906×10^{-02}
2	1	91001.0	2.467×10^{15}	9.324×10^{05}	1.694×10^{-03}
3	2	280090.0	4.569×10^{14}	3.030×10^{05}	9.655×10^{-04}
4	3	563952.0	1.599×10^{14}	1.505×10^{05}	6.805×10^{-04}

Oscillation Neutrinos

Oscillation neutrinos are emitted when parent torices shift from the higher oscillation levels N_k to the lower levels N_j. If b_{1k} and b_{1j} are the spiral radii and m_{ok} and m_{oj} are the rest masses of the respective parent torices then the spiral frequency of the emitted oscillation neutrino f_3 is equal to:

$$f_3 = \frac{C^2}{h}\left(\frac{m_{ok}}{b_{1k}} - \frac{m_{oj}}{b_{1j}} \right) \qquad (G2)$$

Table G5 shows the parameters of the fast oscillation neutrinos that are emitted by a negative etheron as a result of the shift of its inner complementary-polarized torices $-A_{Nm}$ and $-\bar{A}_{Nn}$, located at the excitation level $n = 1$, from the higher oscillation level N_k to the lower level N_j. It is worth to mention that there is a problem of transfer of the inner complementary-polarized torices $-A_{Nm}$ and $-\bar{A}_{Nn}$, located at the excitation level $n = 1$, from the oscillation level $N = 3$ to the level $N = 2$, as will be explained later.

Table G5 Parameters of fast oscillation neutrinos associated with inner complementary-polarized torices $-A_{Nm}$ and $-\overline{A}_{Nn}$ of negative etherons ($n = 1$, $b_1 = 1.00002663$).

Oscillation levels		Parameters of fast oscillation neutrinos ($V_{3t} \sim C$)			
N_k	N_j	b_3	f_3, Hz	m_{hg}, MeV/c^2	Neutrino
1	0	1.00367550	1.236×10^{20}	0.001871	e-neutrino
2	1	1.00183100	1.236×10^{20}	0.001401	e-neutrino
3	2	> 1.890	2.503×10^{22}	24.258	μ-neutrino
4	3	1.06700100	4.097×10^{23}	55.792	τ-neutrino

Table G6 Parameters of fast oscillation neutrinos associated with outer complementary-polarized torices $-A_{Nn}$ and $-\overline{A}_{Nn}$ of a negative lepton ($n = 1$, $b_1 = 37559.724$).

Oscillation levels		Parameters of fast oscillation neutrinos ($V_{3t} \sim C$)			
N_k	N_j	b_3	f_3, Hz	m_{hg}, MeV/c^2	Neutrino
1	0	1.0000000971	3.290×10^{15}	4.964×10^{-08}	e-neutrino
2	1	1.0000000486	3.290×10^{15}	3.723×10^{-08}	e-neutrino
3	2	1.0000065590	6.663×10^{17}	3.465×10^{-04}	μ-neutrino
4	3	1.0000015669	1.091×10^{19}	1.392×10^{-03}	τ-neutrino

Table G7 Parameters of fast oscillation neutrinos associated with complementary-polarized torices $-A_{Nn}$ and $-\overline{A}_{Nn}$ of a negative lepton ($n = 0$, $b_1 = 2.0$).

Oscillation levels		Parameters of fast oscillation neutrinos ($V_{3t} \sim C$)			
N_k	N_j	b_3	f_3, Hz	m_{hg}, MeV/c^2	Neutrino
1	0	1.0000000971	6.178×10^{19}	0.000934	e-neutrino
2	1	1.0000000486	6.178×10^{19}	0.000700	e-neutrino
3	2	1.0000065590	1.251×10^{22}	7.6012	μ-neutrino
4	3	1.0000015669	2.048×10^{23}	26.963	τ-neutrino

Table G6 shows the parameters of the fast oscillation neutrinos that are emitted by a negative lepton as a result of the shift of its outer complementary-polarized torices $-A_{Nm}$ and $-\bar{A}_{Nn}$, located at the excitation level $n = 1$, from the higher oscillation level N_k to the lower level N_j. Table G7 shows the parameters of the fast oscillation neutrinos that are emitted by a negative lepton as a result of the shift of its complementary-polarized torices $-A_{Nm}$ and $-\bar{A}_{Nn}$, located at the excitation level $n = 0$, from the higher oscillation level N_k to the lower level N_j.

Transmission of Photons

One can relate the process of the transmission of photons to the well-known process of the transmission of electromagnetic signals. To ensure that transmitted signals are free from unrelated distortions known as "noise," we usually install intermediate stations to filter the "noise," and then send the clear signal towards the next receiving station.

During the transmissions of photons, the role of intermediate stations is fulfilled by etherons. Fig. G2 shows how this works: after absorbing an incoming photon, the etheron increases its energy level. During its return to the initial energy level, the etheron emits a new photon. This process of the accumulation and release of energy by the etheron is somewhat similar to the process of compression and decompression of air that takes place during the transmission of sound waves.

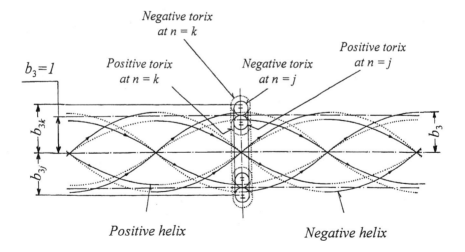

Fig. G2 Transmission of a photon by an etheron.

Let us consider a case in which an etheron is excited by a fast photon. After its excitation, the etheron emits both fast and slow photons of the same frequency. There are two important consequences of this phenomenon. Firstly, since the energies of both fast and slow photons are equal to each other, one half of the energy transmitted by fast photon will be lost during its passage through etherons. This may explain why the intensity of light follows the inverse square law. Secondly, since slow photons have been emitted by etherons since the beginning of time, they have probably by now permeated the entire space, creating *microwave background radiation*. Another possible function of slow photons is the transmission of heat.

Etheron Field

An etheron field is an intermediate stage of the Micro-Big-Bang process. In the beginning of this process, all the torices were located at the highest oscillation level $N = 27$ and the highest excitation level $n \to \infty$. The real torices were reduced to the Zero-Point-Energy (ZPE) rings with the relative spiral radius $b_1 \to 1$, while the imaginary torices were reduced to Maximum-Point-Energy (MPE) rings with the relative spiral radius $b_1 \to 0$. Importantly, at this stage, both the real and imaginary torices had the smallest inversion radius. The ZPE and MPE rings were angularly spaced in respect to each other, forming the *primordial spherix*.

During the Micro-Big-Bang process, the etherons were produced as a result of polarizarion of the ZPE and MPE rings and their subsequent inflation caused by the reduction of their oscillation and excitation levels. The inflation process continued without interruption until after the oscillation level was reduced to $N = 3$. As was shown in Table F7 of Section F, this oscillation level corresponds to muon-based etherons. The rest mass of these etherons is 1.29×10^9 times smaller and their inversion radius is greater by the same ratio in comparison with the respective parameters of the etheron at the beginning of the Micro-Big-Bang inflation process.

Here is the reason why the inflation process was suspended at oscillation level $N = 3$. As we discussed earlier, the shift of the torices from higher to lower oscillation levels is accompanied by the emission of helices that form neutrinos. The emission process provides an avenue for the release of torix energy. This release proceeds expeditiously as long as the energy that can be carried by neutrinos is equal to or greater than the energy that must be released during the shift of torices from higher to lower oscillation levels.

This condition is met during the shift of torices between all the adjacent oscillation levels, except when they have to be shifted from the oscillation level $N = 3$ to $N = 2$. This phenomenon is ilustrated in Fig. G3 in terms of relative helix spiral frequency δ_3 that is proportional to helix energy. The bell-type curve represents the general relationship between relative helix spiral frequency δ_3 and its relative spiral radius b_3. The straight dotted lines represent relative helix spiral frequencies of the helices that would be emitted during the shift of the torices between adjacent oscillation levels.

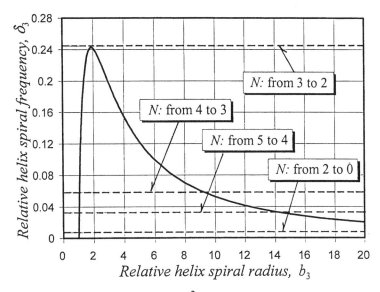

Fig. G3 Relative helix spiral frequency δ_3 as a function of the relative helix spiral radius b_3 (solid line) and the relative spiral frequencies δ_3 of the helices emitted during oscillation of the torices (dotted lines).

The oscillation process continues as long as the relative helix spiral frequencies shown with dotted lines are less than the maximum possible emitted relative helix spiral frequency $\delta_3 = f_3/f_0 = 0.243$ that corresponds to the pick of the bell-type curve. When this condition is met, the dotted lines will have at least one and generally two cross-over points with the bell-type curve that determine both the frequencies and the radii of fast and slow helices. This condition, however, is not met when imaginary torices must be shifted from the oscillation level $N = 3$ to $N = 2$. In that case, the dotted lines are located just slightly above the bell-type curve, with the deficiency of helix energy of only 1.4%. This deficiency, however, is sufficient to suspend the oscillation process.

Subsequently, one may expect the accumulation of muon-type etherons, which would then form the *etheron field*. Like the Higgs field[1], proposed by

Peter Higgs in 1964, the etheron field serves as a pool of matter for the creation of elementary particles. Remarkably, as was shown in Section F, the cores of nucleons are made of muon-based torices located at the oscillation level $N = 3$. It is a different case with the neutron shells that are made of $3e$-based torices ($N = 2$) and e-based torices ($N = 0$). Therefore, to create neutron shells, the torices must be shifted to the oscillation levels N that are lower than 3. Most likely, this shift occurs spontaneously due to external disturbances, because of the small deficiency of helix energy that has to be supplemented.

Nuclear Reactions

According to the SFT, nuclear reactions occur as a result of oscillation and excitation of torices that comprise the elementary particles. The oscillation and excitation processes are accompanied by the emission or absorption of helices that form either photons or neutrinos. As an example, let us consider the nuclear reaction that involves the decay of a neutron into a proton and an electron. During the decay, the core of the neutron remains unchanged and, therefore, the decay is only due to the changes that occur in the neutron shell.

As was shown in Table F10 of Section F, a neutron shell comprises two sets of torices. One set comprises negative real and imaginary torices $-A_{00}$ and $-\bar{A}_{00}$, and the other set comprises positive real and imaginary torices $+A_{20}$ and $+\bar{A}_{20}$. Negative torices are located at oscillation level $N = 0$, while positive torices are located at $N = 2$. All the torices are located at the universal excitation level $n = 0$. Neutron decay involves four transformations of real and imaginary torices (Table G8):

1. Intermediate shift of the positive real and imaginary torices $+A_{20}$ and $+\bar{A}_{20}$ from the oscillation level $N = 2$ to $N = 1$, while maintaining the same universal excitation level $n = 0$. As a result of this shift, they transform respectively into the positive real and imaginary torices $+A_{10}$ and $+\bar{A}_{10}$.
2. Energy released during the intermediate shift of the positive real and imaginary torices $+A_{20}$ and $+\bar{A}_{20}$ from the oscillation level $N = 2$ to $N = 1$ is used to shift the negative real and imaginary torices $-A_{00}$ and $-\bar{A}_{00}$ from the universal excitation level $n = 0$ to $n = \infty$, producing the negative real and imaginary torices $-A_{0\infty}$ and and $-\bar{A}_{0\infty}$ that form a free electron σ^-. The excitation energy level of the electron may eventually reduce to the ground level $n = 1$ at which it will contain the negative real and imaginary torices $-A_{01}$ and and $-\bar{A}_{01}$.

3. Final shift of the positive real and imaginary torices $+A_{10}$ and $+\overline{A}_{10}$ from the oscillation level $N = 1$ to $N = 0$, while maintaining the same universal excitation level $n = 0$. As a result of this shift, they transform respectively into the positive real and imaginary torices $+A_{00}$ and $+\overline{A}_{00}$, forming a proton shell.

4. Energy released during the final shift of the positive real and imaginary torices $+A_{10}$ and $+\overline{A}_{10}$ from the oscillation level $N = 1$ to $N = 0$ is used to produce an *e*-neutrino ν_e.

The equation below describes the nuclear reaction[2]:

$$n^o \rightarrow p^+ + e^- + \nu_e \tag{G3}$$

Table G8 Neutron decay.

Structure before decay				Intermediate stage				Structure after decay			
Particle	Torix	N	n	Particle	Torix	N	n	Particle	Torix	N	n
Negative neutron shell	$-A_{00}$	0	0	Free electron	$-A_{0\infty}$	0	∞	Free electron	$-A_{01}$	0	1
	$-\overline{A}_{00}$	0	0		$-\overline{A}_{00}$	0	∞		$-\overline{A}_{01}$	0	1
Positive neutron shell	$+A_{20}$	2	0	Positive neutron shell	$+A_{10}$	1	0	Proton shell	$+A_{00}$	0	1
	$+\overline{A}_{20}$	2	0		$+\overline{A}_{10}$	1	0		$+\overline{A}_{00}$	0	1

REFERENCES

1. Gribbin, J., *Q is For Quantum, An Encyclopedia of Particle Physics* A Touchstone Book, Simon & Schuster, New York, 2000.
1. Serway, R.A., *Physics For Scientists & Engineers with Modern Physics*, Third Edition, Saunders Golden Sunburst Series, Philadelphia, PA, 1992, p. 1424.

H

THE UNIFICATION OF
STRONG AND ELECTRIC FORCES

Introduction

The electric, magnetic, and gravitational forces are commonly thought to follow the inverse-square law. Rudjer Bošković was the first scientist who questioned this law, as early as 1758. As an alternative, he proposed the universal force law[1]. According to Bošković (Fig. H1.0), the inverse-square law remains valid only when the distance between particles is very large (zone C). As the distance decreases (zone B), the force between the particles ceases to follow the inverse-square law and becomes alternately attractive or repulsive, depending on the distance by which particles are separated. Then at even shorter distances (zone A), the force becomes purely repulsive. As the distance diminishes to zero, repulsion grows infinite, thus preventing direct contact between particles.

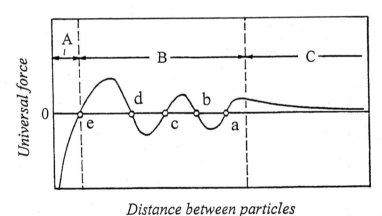

Distance between particles

Fig. H1.0 Graphic presentation of Bošković's Universal Force Law.

According to current theories of particle physics, nucleons are held together by one of the fundamental forces of nature, called the *strong force*. The strong force is very short-ranged and is negligible for separations greater than about 10^{-15} m. The strong force is thought to be mediated by field particles called gluons[2]. It is approximately 100 times stronger than the electromagnetic force and does not follow the inverse-square law[3]. Neutrons are deflected when they pass in the neighborhood of protons[4]. It was also found that when two nucleons get very close to one another, the nucleon-nucleon force becomes strongly repulsive[5].

Some limitations of the current models arise from the presentation of elementary particles as point-like spheres. Alternative model, called *the spinning charge ring model*[6] was proposed by Parson in 1915. In his model, the electron appears as a very thin ring on which the negative charge revolves at a velocity of approximately the velocity of light. In the early 1960s, Penrose proposed the concept of a spinning and propelling object called the *twistor* to build the idea of quantum space[7]. In 1962, Wheeler introduced the idea of a *geon* as a standing electromagnetic wave, or a beam of light, bent into a closed circular toroid of high energy concentration[8]. In 1966, Bostick outlined the electron as a string-like submicroscopic force-free plasmoid constructed by the self-energy of electric and magnetic vectors[9,10]. The spinning charge ring model of the electron was further developed in 1990 by Bergman and Wisely[11], and, in 1996-1998, by Joseph and Charles Lucas[12-14].

In 1996-2001, the author introduced the spiral field theory according to which the elementary particles were presented as assemblies of double-toroidal spiral fields called *torices*, while the electromagnetic waves and neutrinos were presented in the form of double-helical spiral fields called *helices*[15-20]. The same concept was applied to both nucleons, proton and neutron. They were also assumed to be formed from torices. Cores of neutron and proton are identical and are made of positive and negative torices located at the third oscillation level. The nucleon shells, however, are different. The neutron shell is made of inner positive and outer negative torices located respectively at the second and zero oscillation levels. The proton shell is made of only positive torices located at the zero oscillation level. The strong forces are mainly dependent on the interaction between nucleon shells.

The strong forces between nucleons are calculated by replacing the torices with electric mini charges evenly distributed along the torix circular orbital paths, and by the subsequent application of Coulomb's law. The calculations show that both the sign and the magnitude of electric forces between nucleons changes sharply as the distance between them becomes comparable with their radii. Both the attraction and the repulsion forces increase steeply and reach their maximum values as the distances between the nucleons reduce and become with the nucleon radius. At these distances, the strong forces do not follow the inverse-square law. At distances that are more then five times greater then the nucleon radius, the attraction force between a proton and a neutron decreases inversely proportional to the fourth power of the distance between them. The repulsion force between two neutrons decreases inversely proportional to the sixth power, while the repulsion force between two protons decreases inversely proportional to the second power of the distance between them.   The proposed theory does not require an additional particle to explain the strong interaction, leaving this function to the proton and the neutron. Also, there is no need for a new fundamental force of nature, because the strong interaction can be presented as a particular manifestation of the electric force. The proposed theory explains the role of neutrons in holding the nucleons together in the nuclei. It makes clear a cause for the *nucleon scattering*. Finally, it explains a possibility of existence of neutron clusters, and particularly, the *tetraneutrons*.

H1. Electric Forces Between Circularly-Distributed Charges

Let us consider two sets of electric charges representing two adjacent particles separated by a distance S. The electric mini charges in each set are distributed along two concentric circles as shown in Fig. H1.1.

In one set, the electric mini charges are distributed along the inner circle with the relative radius b_a and along the outer circle with the radius b_A. In this set, the angular positions of the electric mini charges along the circles are determined by the angles α_j. In the other set, the electric mini charges are distributed along the inner circle with the radius b_b and along the outer circle with the relative radius b_B. In this set, the angular positions of the electric mini charges along the circles are determined by the angles β_k.

According to Coulomb's law, electric force $F(ab)$, applied to electric mini charges e_a and e_b located at points a and b and separated by the distance ab, is equal to:

$$F(ab) = - \frac{ke_a e_b}{(ab)^2} \qquad \text{(H1-1)}$$

where
   k = Coulomb constant.

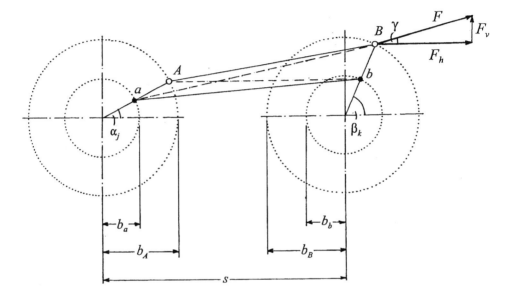

Fig. H1.1 Distributed electric mini charges.

Similarly, the electric forces $F(aB)$, $F(AB)$, and $F(Ab)$ are equal to:

$$F(aB) = - \frac{ke_a e_B}{(aB)^2} \qquad \text{(H1-2)}$$

$$F(AB) = - \frac{ke_A e_B}{(AB)^2} \qquad \text{(H1-3)}$$

$$F(Ab) = -\frac{ke_A e_b}{(Ab)^2} \qquad \text{(H1-4)}$$

It is convenient to express the geometrical parameters in relative terms in respect to the torix inversion radius r_i that is equal to the nucleon outer radius and is given by: .

$$r_i = \frac{ke_o^2}{2m_o C^2} = 1.408971\times10^{-15}\,m \qquad \text{(H1-5)}$$

where

e_o = electron rest electric charge
m_{oe} = electron rest mass
C = ultimate spiral field velocity.

Thus, the relative geometrical parameters are equal to:

$$b_a = \frac{r_a}{r_i}; \qquad b_A = \frac{r_A}{r_i}; \qquad b_b = \frac{r_b}{r_i}; \qquad b_B = \frac{r_B}{r_i}; \qquad s = \frac{S}{r_i}. \qquad \text{(H1-6)}$$

The electric mini charges are equal to:

$$e_a = \frac{e_o}{n_e}\epsilon_a; \qquad e_A = \frac{e_o}{n_e}\epsilon_A; \qquad e_b = \frac{e_o}{n_e}\epsilon_b; \qquad e_B = \frac{e_o}{n_e}\epsilon_B \qquad \text{(H1-7)}$$

where

ϵ_a, ϵ_A, ϵ_b, and ϵ_B are the torix relative electric charges distributed along the circles with radii r_a, r_A, r_b, and r_B respectively
n_e = the number of the electric mini charges distributed along the perimeter of each circle.

The relative horizontal components of electric forces acting between electric charges located at points a, b, A and B (Fig. H1.1) are equal to:

$$f_h(ab) = F_h(ab)\frac{r_i^2}{ke_o^2}$$ (H1-8)

$$f_h(aB) = F_h(aB)\frac{r_i^2}{ke_o^2}$$ (H1-9)

$$f_h(AB) = F_h(AB)\frac{r_i^2}{ke_o^2}$$ (H1-10)

$$f_h(Ab) = F_h(Ab)\frac{r_i^2}{ke_o^2}$$ (H1-11)

From Eqs. (H1-1) - (H1-11), we obtain the final equations for the relative horizontal components of electric forces:

$$f_h(ab) = -\frac{\epsilon_a\epsilon_b}{n_e^2}\frac{s+b_b\cos\beta_k-b_a\cos\alpha_j}{\left((s+b_b\cos\beta_k-b_a\cos\alpha_j)^2+(b_b\sin\beta_k-b_a\sin\alpha_j)^2\right)^{3/2}}$$ (H1-12)

$$f_h(aB) = -\frac{\epsilon_a\epsilon_B}{n_e^2}\frac{s+b_B\cos\beta_k-b_a\cos\alpha_j}{\left((s+b_B\cos\beta_k-b_a\cos\alpha_j)^2+(b_B\sin\beta_k-b_a\sin\alpha_j)^2\right)^{3/2}}$$ (H1-13)

$$f_h(AB) = -\frac{\epsilon_A \epsilon_B}{n_e^2} \frac{s + b_B \cos\beta_k - b_A \cos\alpha_j}{\left((s + b_B \cos\beta_k - b_A \cos\alpha_j)^2 + (b_B \sin\beta_k - b_A \sin\alpha_j)^2\right)^{3/2}} \quad \text{(H1-14)}$$

$$f_h(Ab) = -\frac{\epsilon_A \epsilon_b}{n_e^2} \frac{s + b_b \cos\beta_k - b_A \cos\alpha_j}{\left((s + b_b \cos\beta_k - b_A \cos\alpha_j)^2 + (b_b \sin\beta_k - b_A \sin\alpha_j)^2\right)^{3/2}} \quad \text{(H1-15)}$$

H2. The Electric Interactions Between Nucleons

Figs. H2.1- H2.3 show the tori that form the shells of two interacting nucleons. Three types of nucleon-nucleon interactions are considered: proton-proton (*p-p*), proton-neutron (*p-n*), and neutron-neutron (*n-n*).

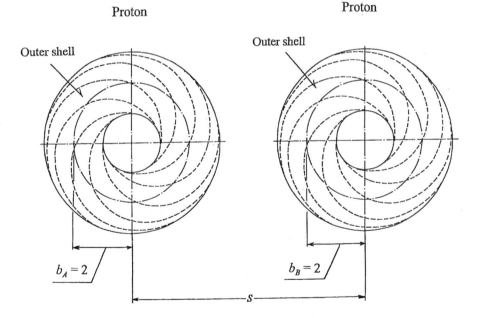

Fig. H2.1 Shells of two adjacent protons.

When considering the interaction between these nucleons, the orbital paths of torices are replaced with electric mini charges distributed along the circles with the relative radii b_A, b_a, b_B and b_b. Table H2.1 shows the values of the relative radii b_A, b_a, b_B and b_b and the values of the respective torix relative electric charges ϵ_A, ϵ_a, ϵ_B and ϵ_b.

Table H2.1 Torix parameters of interacting nucleons.

Inter-action	Torix parameters							
	Relative spiral radius				Relative electric charge			
	b_a	b_A	b_b	b_B	ϵ_a	e_A	ϵ_b	ϵ_B
p-p	0	2	0	2	0	+1	0	+1
p-n	0	2	2/3	2	0	+1	+2/3	-1
n-n	2/3	2	2/3	2	+2/3	-1	+2/3	-1

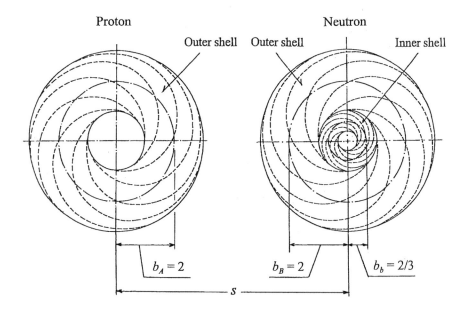

Fig. H2.2 Shells of adjacent proton and neutron.

As an example, Figs. H2.4 - H2.6 show graphs of the relative horizontal components of electric forces f_h between electric mini charges of two nucleon shells separated by a relative distance $s = 4.0$, as calculated from the equations shown below. Plots shown in these figures are given for selected angles $\beta_k = 62.5, 72.5, 82.5, 92.5, 102.5,$ and $112.5°$ when the angles α_j cover the range from 0 to 360°. The number of electric mini charges in each circle n_e is equal to 72.

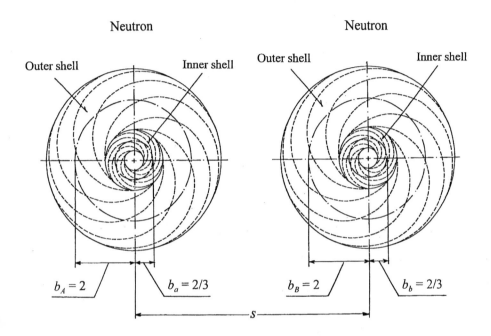

Fig. H2.3 Shells of two adjacent neutrons.

For the $(p$-$p)$-interaction, the relative horizontal components of the electric forces f_h is equal to:

$$f_h = f_h(ab)$$ (H2-1)

For a $(p$-$n)$-interaction, the relative horizontal component of electric forces f_h is equal to:

$$f_h = f_h(ab) + f_h(aB)$$ (H2-2)

The Unification of Strong and Electric Forces

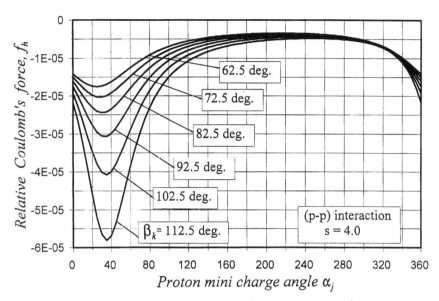

Fig. H2.4 Relative horizontal component of Coulomb's force f_h between circularly-distributed electric mini charges representing two proton shells.

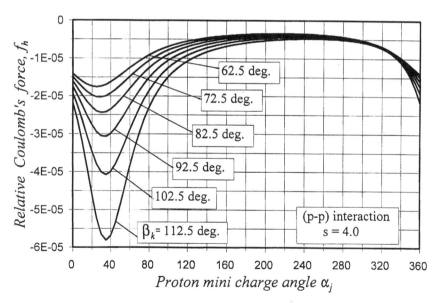

Fig. H2.5 Relative horizontal component of Coulomb's force f_h between circularly- distributed electric mini charges representing the shells of a proton and a neutron.

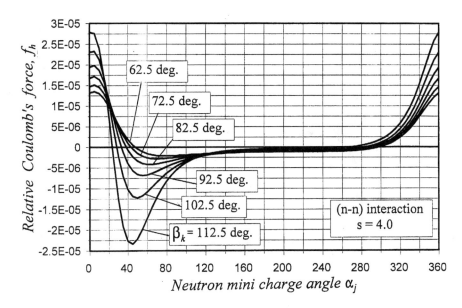

Fig. H2.6 Relative horizontal component of Coulomb's force f_h between circularly-distributed electric mini charges representing two neutron shells.

For a $(n\text{-}n)$-interaction, the relative horizontal component of electric forces f_h is equal to:

$$f_h = f_h(ab) + f_h(aB) + f_h(AB) + f_h(Ab) \tag{H2-3}$$

H3. Strong Forces Between Nucleons

By definition, the strong force is the electric force applied to two adjacent nucleons separated by a distances comparable with the nucleon dimensions. To calculate the relative strong force f_s, it is necessary to sum up the relative horizontal components of Coulomb's forces f_h between all the electric mini charges contained in both nucleons. Thus, we obtain:

$$f_s = \sum f_h \qquad (0 \le \alpha_j < 360^o \ \wedge \ 0 < \beta_k \le 360^o) \tag{H3-1}$$

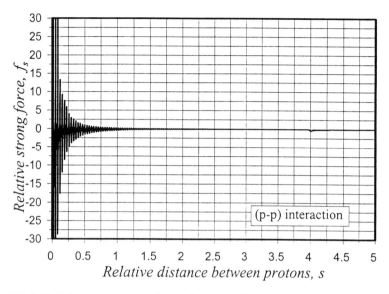

Fig. H3.1 Relative strong force f_s as a function of relative distance between two adjacent protons s $(0 \leq s \leq 5)$.

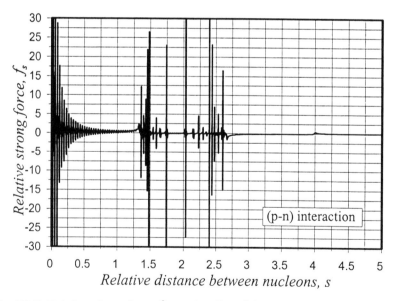

Fig. H3.2 Relative strong force f_s as a function of the relative distance between proton and neutron s $(0 \leq s \leq 5)$.

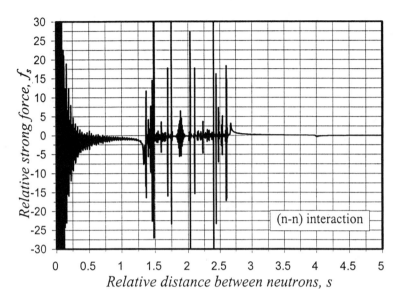

Fig. H3.3 Relative strong force f_s as a function of the relative distance between two adjacent neutrons s $(0 \leq s \leq 5)$.

In the calculations of the relative strong forces f_s shown in Figs. H3.1-H3.13, the number of electric mini charges in each circle n_e is 360, and the angles α_j and β_k vary from 0 to 359° and from 0.5 to 359.5° respectively. The increments of the relative distance between nucleons s is 0.001. In Figs H3.1 - H3.3, the selected range of relative distance between the nucleons s is from 0 to 5. There are three distinct subranges of s within this range. Within the first subrange, s changes from 0 to 1. Within the second subrange, s changes from 1 to 3, and within the third subrange, s changes from 3 to 5.

The first subrange - The variation of the relative strong force f_s within the first subrange is shown in Figs. H3.4-H3.6. For all three types of nucleon interactions, the relative strong forces f_s oscillates continuously without any interruption, and the amplitude of the oscillations increases as the relative distance between nucleons s decreases. The sign of the average strong force f_s depends of the type of the nucleon interaction. For the (p-p) interaction, f_s tends to be negative (repulsion force); for the (p-n) interaction, f_s tends to be positive (attraction force), while for the (n-n) interaction, it is distinctly negative. As the relative distance between nucleons s decreases, the oscillations become more symmetrical in respect to zero. Shown on the graph related to the (p-p) and (p-n) interactions (Figs. H3.4 and H3.5) are the points *a* corresponding to the transitions from the attraction to repulsion strong forces. These points indicate the distances between the nucleons at which they may become interlocked.

The second subrange - The variation of the relative strong force f_s within the second subrange is shown in Figs. H3.7 - H3.9. For the (p-p) interaction, the relative strong force f_s oscillates continuously without interruption similarly to the first subrange. In that case, the protons may become interlocked in points corresponding to the transitions from the attraction to repulsion strong forces. For the (p-n) and (n-n) interactions, however, the relative strong force f_s behaves in a completely different way. It oscillates intermittently and has only a few oscillation spikes. Notice that in the (p-n) interactions, a negative spike precedes a positive spike as the distance between nucleons decreases (Fig. H3.8). Therefore, the nucleons may become interlocked in the zones located between the adjacent positive spikes. Reversed sequence of spikes occurs in the (n-n) interactions (Fig. H3.9). In that case, the nucleons may become interlocked in the points of transition from attraction to repulsion forces.

The third subrange - The variation of the relative strong force f_s within the third subrange is shown in Figs. H3.10 - H3.12. Within this ranges, for all three types of interaction, there is only one change of the sign of the relative strong force f_s. This force reaches its maximum absolute value as the relative distance between the nucleons *s* decreases to about 4. The absolute value of this force then sharply decreases as the relative distance between the nucleons *s* becomes less than 4.

Notice that within the third subrange the maximum absolute value of the strong force f_s is significantly smaller than that within the first and second subranges. Shown on the graph related to the (p-n) interaction (Fig. H3.11) is the point *a* corresponding to a transition from the attraction to repulsion strong forces. This point indicates the largest distance between the proton and neutron at which they may become interlocked.

For all three types of nucleon-nucleon interactions, when relative distances between the nucleons $s > 5$ the relationship between relative strong force f_s and relative distance between nucleons *s* can be presented with an exponential equation:

$$f_s = \frac{A}{s^x} \qquad \text{(H3-2)}$$

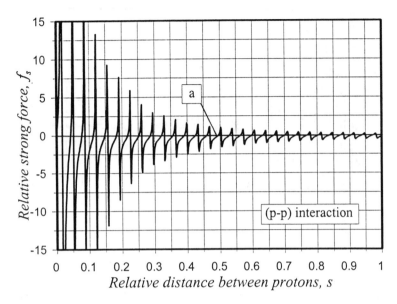

Fig. H3.4 Relative strong force f_s as a function of relative distance between two adjacent protons s within the first subrange $(0 \leq s \leq 1)$.

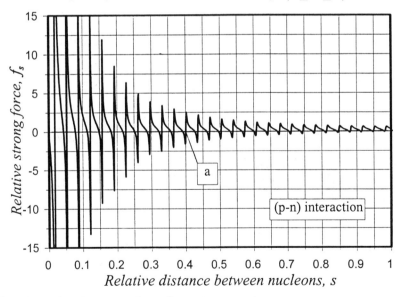

Fig. H3.5 Relative strong force f_s as a function of the relative distance between proton and neutron s within the first subrange $(0 \leq s \leq 1)$.

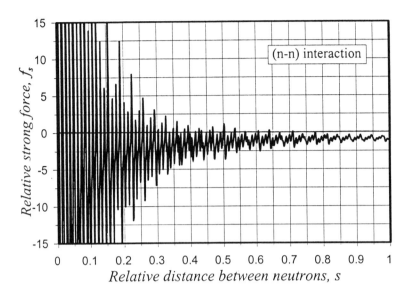

Fig. H3.6 Relative strong force f_s as a function of the relative distance between two adjacent neutrons s within the first subrange $(0 \leq s \leq 1)$.

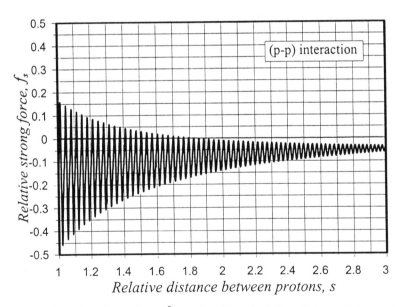

Fig. H3.7 Relative strong force f_s as a function of relative distance between two adjacent protons s within the second subrange $(1 \leq s \leq 3)$.

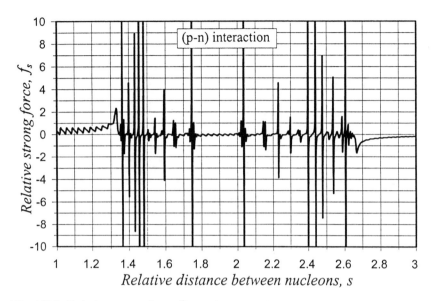

Fig. H3.8 Relative strong force f_s as a function of the relative distance between proton and neutron s within the second subrange ($1 \leq s \leq 3$).

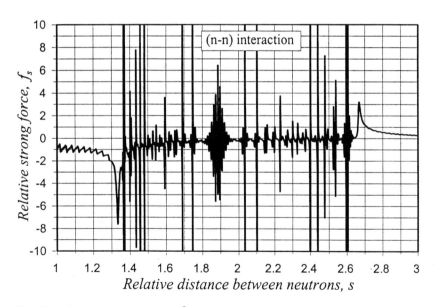

Fig. H3.9 Relative strong force f_s as a function of the relative distance between two adjacent neutrons s within the second subrange ($1 \leq s \leq 3$).

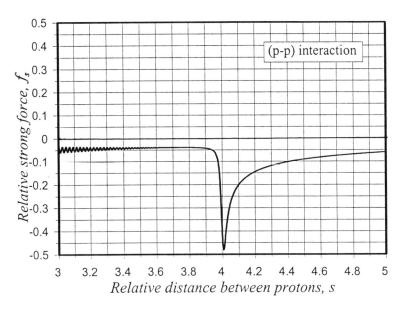

Fig. H3.10 Relative strong force f_s as a function of relative distance between two adjacent protons s within the third subrange $(3 \leq s \leq 5)$.

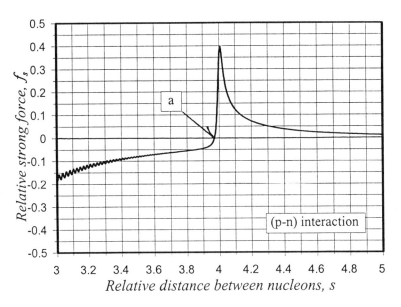

Fig. H3.11 Relative strong force f_s as a function of the relative distance between proton and neutron s within the third subrange $(3 \leq s \leq 5)$.

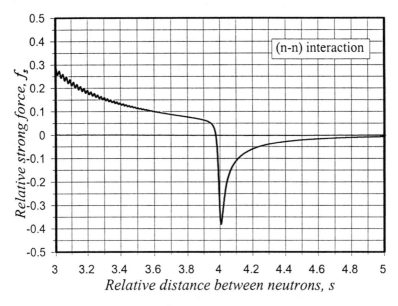

Fig. H3.12 Relative strong force f_s as a function of the relative distance between two adjacent neutrons s within the third subrange $(3 \leq s \leq 5)$.

Table H3.1 gives the values of the exponents x and the approximate values of the coefficients A of Eq. (H3-2) for three types of nucleon-nucleon interactions for various ranges of the relative distance between nucleons s.

Table H3.1 Parameters of the strong force equation (H3-2).

Type of interaction		Range of s				
		5 - 500	10 - 500	20 - 500	40 -500	100 -500
(p-p)	A	-157.412	-49.441	-39.109	-36.635	-35.816
	x	6.2394	6.0530	6.0153	6.0048	6.0012
(p-n)	A	+11.283	+3.686	+2.929	+2.746	+2.686
	x	4.2321	4.0521	4.0141	4.0047	4.0011
(n-n)	A	-1.694	-1.110	-1.030	-1.009	-1.002
	x	2.0848	2.0168	2.0048	2.0015	2.0004

Remarkably, as follows from Table H3.1, when the relative distance between nucleons $s > 5$, the strong force between them decreases inverse proportionally to the distance with the exponent x approaching to:

For (p-p) strong interaction: $x \to 2.0$
For (p-n) strong interaction: $x \to 4.0$
For (n-n) strong interaction: $x \to 6.0$.

Fig. H3.13 shows a comparison of relative strong force f_s for three types of strong interaction as a function of relative distance between two adjacent nucleon s.

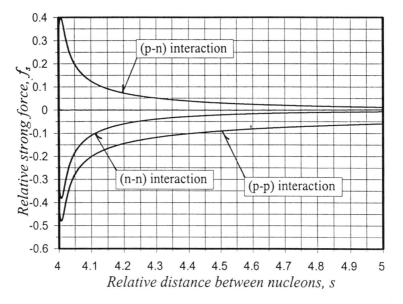

Fig. H3.13 Comparison of relative strong force f_s for three types of strong interaction as a function of relative distance between two adjacent nucleon s
$$(4 \leq s \leq 5).$$

H4. Formation of nuclei and particle clusters

As we mentioned earlier, the proton and neutron may become interlocked within the third subrange of the relative distance between nucleons s (Figs. H3.10 - H3.12). As an example, Fig. H4.1 shows a structure containing two protons p and two neutrons n. The protons are separated from neutrons by the relative distance $s = 4$. The protons are separated from each other by the distance $s = 5.567$. The neutrons are separated from each other by the same distance as the protons. Importantly, at these distances, the repulsion forces between protons and between neutrons are substantially smaller than the maximum attraction forces between protons and neutrons.

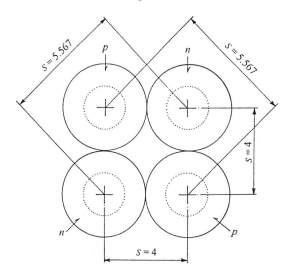

Fig. H4.1 (2p + 2n) structure within the third subrange.

As we discussed earlier, the nucleons can also become interlocked within the first and second subranges of the relative distance between nucleons s. One of the possible structures corresponding to the second subrange is a neutron cluster containing four neutrons that is known as the *tetraneutron*[21]. By applying the above described method of the unification of strong and electric forces, it is also possible to explain formation of the *high-density electron clusters*[22].

REFERENCES

1. Boscovich, R. J., *A Theory of Natural Philosophy*, English Edition, The M.I.T. Press, Cambridge, MA, 1966.

2. Serway, R. A., *Physics for Scientists & Engineers with Modern Physics*, Third Edition, Saunders Golden Sunburst Series, Saunders College Publishing, Philadelphia, 1992.

3. Gribbin, J., *Q is For Quantum - An Encyclopedia of Particle Physics*,

Simon & Schuster, New York, 1998.

4. Born, M., *Atomic Physics*, Dover Publication, Inc., New York, 1969.

5. Boorse, H. A. and Motz, L., *The World of the Atom*, Basic Books, Inc., New York, 1966, pp. 1619-24.

6. Parson, *Smithsonian Miscellaneous Collections*, Vol. 65, No. 11, 1915.

7. Hawking, S. and Penrose, R., *The Nature of Space and Time*, Princeton University Press, Princeton, N. J., 1966.

8. Wheeler, J. A., *Geometrodynamics, Topics of Moder Physics*, Vol. 1, Academic Press, New York, 1962.

9. Bostick, W. H., *Physics of Fluids*, Vol. 9, p. 2079, (1966).

10. Bostick, W. H., "Mass, Charge, and Current: The Essence and Morphology," *Physics Essays*, Vol. 4, No. 1, pp. 45-59, 1991.

11. Bergman, D. L. and Wisely, P., " Spinning Charged Ring Model of Electron Yielding Anomalous Magnetic Moment," *Galilean Electrodynamics*, Vol. 1, No. 5, pp. 63-67, (Sept./Oct. 1990).

12. Lucas, J., " A Physical Model for Atoms and Nuclei," *Galilean Electrodynamics*, Vol. 7, p. 3-12, (1966).

13. Lucas Jr., C. W. and Lucas, J., "A New Foundation for Modern Science," *Proceedings of the International Conference on Creationism*, held at Geneva College in Pennsylvania, USA, Aug. 3-8, 1998.

14. Lucas Jr., C. W., "A New Foundation for Modern Science," *Common Sense Science, Foundations of Science*, Vol. 4, No. 4, Nov. 2001.

15. Ginzburg, V. B., "Toroidal Spiral Field Theory," *Speculations in Science and Technology*, Vol. 19, 1996.

16. Ginzburg, V. B., "Structure of Atoms and Fields," *Speculations in Science and Technology*, Vol. 20, 1997.

17. Ginzburg, V. B., "Double Helical and Double Toroidal Spiral Fields," *Speculations in Science and Technology*, Vol. 22, 1998.

18. Ginzburg, V. B., "Nuclear Implosion," *Journal of New Energy*, Vol. 3, No. 4, 1999.

19. Ginzburg, V. B., "Dynamic Aether," *Journal of New Energy*, Vol. 6, No. 1, 2001.

20. Ginzburg, V. B., *Unified Spiral Nature of the Quantum and Relativistic Universe*, Helicola Press, Pittsburgh, PA, 2002.

21. *New Scientist,* vol. 176, Issue 2336, October 2002, p. 30.

22. Fox, H. and Jin, S.X., "Low-Energy Nuclear Reactions and High-Density Charge Clusters," *Journal of New Energy*, Vol. 3, No. 2/3, 1998.

I

THE UNIFICATION OF
GRAVITATIONAL AND ELECTRIC FORCES

I1. Gravity Versus Electricity

Scientists noticed a long time ago a close correlation between gravitational and electric forces. According to Newton's law of gravitation, the gravitational force F_g between two adjacent bodies is proportional to the product of masses m_1 and m_2 of these bodies and inversely proportional to the square of the distance S between them:

$$F_g = \frac{m_1 m_2 G}{S^2} \qquad \text{(I1-1)}$$

where
$\quad G$ = gravitational constant.

Similarly, according to Coulomb's law, the electric force F_e between two electrically-charged adjacent bodies is proportional to the product of electric charges e_1 and e_2 of these bodies and inversely proportional to the square of the distance S between them:

$$F_e = -\frac{k e_1 e_2}{S^2} \qquad \text{(I1-2)}$$

where
$\quad k$ = Coulomb constant.

In spite of the obvious similarities between the above two equations, there is a principal difference between the forces that they calculate. The gravitational force F_g is always positive (attraction), while the electric force F_e can be either positive (attraction) or negative (repulsion). Scientists proposed numerous ideas to reconcile this difference. As early as in the middle of the eighteenth century, Father Beccaria advanced the fascinating idea that gravity may be explained as the result of a slight deviation of celestial bodies from the state of electric neutrality. P. M. Roget and O. F. Mossotti promoted this idea in the early nineteenth century. Mossoti claimed to describe this idea in mathematical terms using a variant of Poisson's theory. Michael Faraday (1791-1867) was very enthusiastic about Mossotti's work, but have made an attempt to develop his own theory that he called the *theory of electrogravity*[1].

In the 20th century, the idea of the unification of the gravity and electricity was revived as a part of a global task of creating a unified field theory. It was hoped that this theory would unify not only all the forces of the universe but also quantum mechanics with the theory of relativity. Unification of gravity with the other forces was a subject of numerous scientific works[2-13]. Some scientists consider the superstring theory as a likely candidate for the unified field theory[2-5]. This book shows the great potential of the SFT in solving this problem. In Section H of this book we applied the SFT for the unification of strong and electric forces[13]. In this section, we will illustrate how this theory can be applied for the unification of gravitational and electric forces.

I2. Electromass Ratio

According to the SFT, gravitational force depends on two parameters of a body, gravitational mass and electric charge. As follows from Eqs. (I1-1) and (I1-2), when gravitational force F_g is equal to electric attractive force F_e, the ratio of electric charge to gravitational mass is equal to:

$$\Omega = \frac{e_1}{m_1} = \frac{e_2}{m_2} = \sqrt{\frac{G}{k}} \qquad (I2\text{-}1)$$

Since both the gravitational constant G and the Coulomb constant k do not vary, the ratio of electric charge e to gravitational mass m, is constant for all bodies. We will refer to this parameter as *gravitational electromass ratio* Ω.

In spite of the simplicity of derivation of the gravitational electromass ratio Ω, several principal questions remain unanswered. What is the physical meaning of electric charge? What is the sign of the electric charge? We can explain the attractive electric force between two bodies if body 1 is electrically positive while body 2 is negative. But what about a case with three bodies? If the electric charge of body 3 is negative, then it will be attracted by body 1 and repulsed by body 2. Conversely, if the electric charge of body 3 is positive then it will be attracted by body 2 and repulsed by body 1.

Thus, in the case of more than two bodies, the concept of a body having either positive or negative electric charge cannot be applied to explain always attractive character of the gravitational force. SFT finds a solution of this problem by employing the concept of gravitational spiral field.

13. Gravitational Spiral Field of a Single Body

In previous sections of this book, we presented nucleons and electrons that make up the atoms as assemblies of toroidal spiral fields. Since all regular bodies of matter, from grains of sand and stones to planets and galaxies, are made of atoms, it is logical to conclude that these bodies are also assemblies of toroidal spiral fields. These assemblies, however, are so sophisticated that one may rightfully question if it would be possible to describe them in mathematical terms.

Spiral field theory offers a rather simple solution to this problem. It substitutes trillions and trillions of individual toroidal spiral fields that make up a regular body with an integrated toroidal spiral field called the *gravitational spiral field*. Similarly to the toroidal spiral field associated with the particles of the micro-world, the gravitational spiral field comprises two principal components, real and imaginary, with the flow of energy (and time) in opposite directions so that the total energy of the gravitational spiral field remains equal to zero.

Also, as we established for the toroidal spiral fields that reign in the micro-world, both the real and imaginary components of the gravitational spiral field are further polarized into two components, outer negative, or left-handed, and inner positive, or right-handed. Figure I3.1 illustrates the real components of gravitational spiral field, the inner spiral field with spiral radius r_a and the outer spiral field with spiral radius r_A.

Principal relationships in the gravitational spiral field are the same as in the toroidal spiral fields of the micro-world. In both cases, the length of one winding of the trailing inner and outer spirals is equal to the perimeters of respective leading circles with the radius r_a and r_A; also, the spiral field propagates along its spiral path with the ultimate spiral field velocity C.

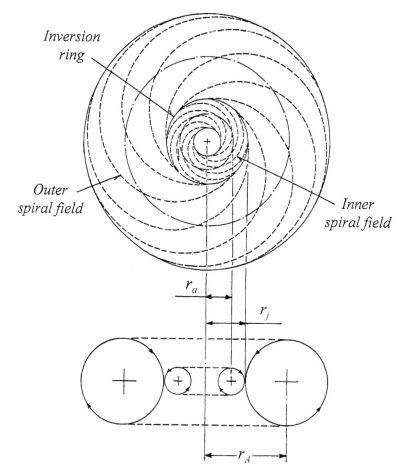

Fig. I3.1 Real component of gravitational spiral field of a body.

The only difference in application to gravity is the definition of the inversion radius. As was shown in the previous sections of this book, for the toroidal spiral field associated with a particle with the rest electric charge e_o and the rest mass m_o, the inversion radius r_i is equal to:

$$r_i = \frac{ke_o^2}{2m_oC^2}$$

(13-1)

For a gravitational field representing a body with gravitational mass m, the inversion radius r_j is equal to[12,13]:

$$r_j = \frac{mG}{2C^2}$$

(13-2)

14. Electric Forces Between Circularly-Distributed Charges

Let us consider two sets of electric charges representing gravitational fields of two adjacent bodies separated by a distance S. The electric mini charges in each set are evenly distributed along two concentric circles as shown in Fig. 14.1.

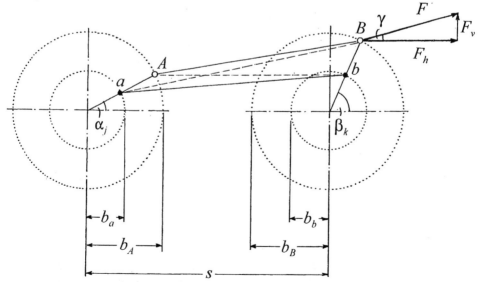

Fig. 14.1 Distributed electric mini charges.

In one set, the electric mini charges are distributed along the inner circle with the relative radius b_a and along the outer circle with the radius b_A. In this set, the angular positions of the electric mini charges along the circles are determined by the angles α_j. In the other set, the electric mini charges are distributed along the inner circle with the radius b_b and along the outer circle with the relative radius b_B. In this set, the angular positions of the electric mini charges along the circles are determined by the angles β_k.

According to Coulomb's law, electric force $F(ab)$, applied to electric mini charges e_a and e_b located at points a and b and separated by the distance ab, is equal to:

$$F(ab) = - \frac{ke_a e_b}{(ab)^2} \tag{I4-1}$$

where
    k = Coulomb constant.

Similarly, the electric forces $F(aB)$, $F(AB)$, and $F(Ab)$ are equal to:

$$F(aB) = - \frac{ke_a e_B}{(aB)^2} \tag{I4-2}$$

$$F(AB) = - \frac{ke_A e_B}{(AB)^2} \tag{I4-3}$$

$$F(Ab) = - \frac{ke_A e_b}{(Ab)^2} \tag{I4-4}$$

It is convenient to express the geometrical parameters in relative terms in respect to the inversion radius r_i of the electron-based toroidal spiral field that is equal to the nucleon outer radius and is given by:

$$r_i = \frac{ke_o^2}{2m_oC^2} = 1.408971 \times 10^{-15} m \qquad (I4\text{-}5)$$

where
e_o = electron rest electric charge
m_o = electron rest mass
C = ultimate spiral field velocity.

Thus, the relative geometrical parameters are equal to:

$$b_a = \frac{r_a}{r_i}; \qquad b_A = \frac{r_A}{r_i}; \qquad b_b = \frac{r_b}{r_i}; \qquad b_B = \frac{r_B}{r_i}; \qquad s = \frac{S}{r_i}. \qquad (I4\text{-}6)$$

Mini electric charges e_a, e_A, e_b, e_B of the gravitational spiral fields of two adjacent bodies 1 and 2 with gravitational masses m_1 and m_2 are equal to:

$$e_a = -e_A = \frac{e_1}{n_e}; \qquad e_b = -e_B = \frac{e_2}{n_e} \qquad (I4\text{-}7)$$

where
e_1 = absolute total value of either electric mini charges e_a or electric mini charge e_A that represent the gravitational spiral field of body 1.
e_2 = absolute total value of either electric mini charges e_b or electric mini charge e_B that represent the gravitational spiral field of body 2.
n_e = the number of the electric mini charges distributed along the perimeter of each circle.

The relative horizontal components of electric forces acting between electric charges located at points a, b, A and B (Fig. I4.1) are equal to:

$$f_h(ab) = F_h(ab) \frac{r_i^2}{ke_o^2} \qquad (I4\text{-}8)$$

$$f_h(aB) = F_h(aB) \frac{r_i^2}{ke_o^2} \tag{I4-9}$$

$$f_h(AB) = F_h(AB) \frac{r_i^2}{ke_o^2} \tag{I4-10}$$

$$f_h(Ab) = F_h(Ab) \frac{r_i^2}{ke_o^2} \tag{I4-11}$$

From Eqs. (I4-1) - (I4-11), we obtain the final equations for the relative horizontal components of electric forces:

$$f_h(ab) = - \frac{e_1 e_2}{(e_o n_e)^2} \frac{s + b_b \cos\beta_k - b_a \cos\alpha_j}{\left((s + b_b \cos\beta_k - b_a \cos\alpha_j)^2 + (b_b \sin\beta_k - b_a \sin\alpha_j)^2\right)^{3/2}} \tag{I4-12}$$

$$f_h(aB) = \frac{e_1 e_2}{(e_o n_e)^2} \frac{s + b_B \cos\beta_k - b_a \cos\alpha_j}{\left((s + b_B \cos\beta_k - b_a \cos\alpha_j)^2 + (b_B \sin\beta_k - b_a \sin\alpha_j)^2\right)^{3/2}} \tag{I4-13}$$

$$f_h(AB) = - \frac{e_1 e_2}{(e_o n_e)^2} \frac{s + b_B \cos\beta_k - b_A \cos\alpha_j}{\left((s + b_B \cos\beta_k - b_A \cos\alpha_j)^2 + (b_B \sin\beta_k - b_A \sin\alpha_j)^2\right)^{3/2}} \tag{I4-14}$$

$$f_h(Ab) = \frac{e_1 e_2}{(e_o n_e)^2} \frac{s + b_b \cos\beta_k - b_A \cos\alpha_j}{\left((s + b_b \cos\beta_k - b_A \cos\alpha_j)^2 + (b_b \sin\beta_k - b_A \sin\alpha_j)^2\right)^{3/2}} \quad \text{(I4-15)}$$

Relative horizontal component of Coulomb's forces f_h between electric mini charges e_a, e_A, e_b and e_B is equal to:

$$f_h = f_h(ab) + f_h(aB) + f_h(AB) + f_h(Ab) \quad \text{(I4-16)}$$

To use the above equations, it is necessary to know the relative geometrical parameters defined by Eq. (I4-6) and mini electric charges defined by Eq. (I4-7). The definition of these parameters will be discussed below.

I5. Gravitational Spiral Fields

Gravitational forces between two adjacent bodies 1 and 2 are defined by interaction between their gravitational spiral fields. Shown in Fig. I5-1 is one pair of the real components of the gravitational spiral fields. According to SFT, the infinite number of spiral fields associated with these two bodies can be substituted with one pair of spiral fields called the *gravitational spiral fields*. The gravitational spiral fields have specific geometrical and electric parameters.

Relative spiral radii of the outer gravitational spiral fields b_A and b_B are equal to the relative distance between bodies s, i.e.:

$$b_A = b_B = s \quad \text{(I5-1)}$$

Relative spiral radii of the inner gravitational spiral fields b_a and b_b are equal to:

$$b_a \approx \frac{r_{j1}}{2r_i}; \quad b_b \approx \frac{r_{j2}}{2r_i} \tag{I5-2}$$

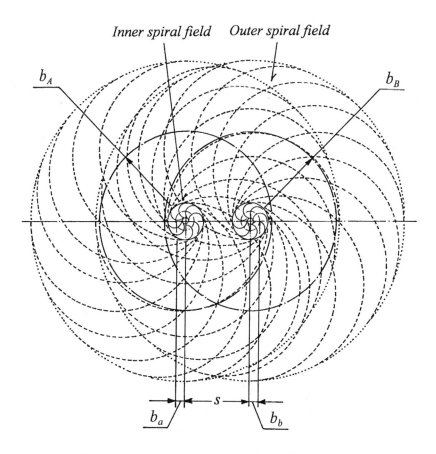

Inner spiral field   Outer spiral field

Fig. I5.1 Real components of one set of spiral fields associated with two adjacent bodies.

Absolute values of electric charges e_1 and e_2 are defined by:

$$e_1 = \frac{\Omega m_1}{n_e \sqrt{\varkappa}}; \quad e_2 = \frac{\Omega m_2}{n_e \sqrt{\varkappa}} \tag{I5-3}$$

x = electric mini charge distribution factor.

The electric mini charge distribution factor x depends on the selected number of electric mini charges distributed along the perimeter of each circle n_e. When n_e approaches infinity, the electric mini charge distribution factor x becomes equal to 3.0.

When we consider the interaction between gravitational spiral fields representing two adjacent bodies, we replace these fields with electric mini charges e_a, e_A, e_b and e_B evenly distributed along the circles with the relative radii b_a, b_A, b_b and b_B. As an example, Fig. I5.2 shows graphs of the relative horizontal components of Coulomb's forces f_h between electric mini charges associated with gravitational spiral fields of two bodies with gravitational masses $m_1 = 10$ kg and $m_1 = 5$ kg; the bodies are separated by the distance $S = 1$m. Plots shown in this figure are given for selected ranges of angles $\beta_k = 90\text{-}108°$, $144\text{-}162°$, and $198\text{-}216°$ when the angles α_j cover the range $\pm 180°$. The number of electric mini charges in each circle n_e were selected to be equal to 7200, and the electric mini charge distribution factor x was assumed to be equal to 1.

Notice three distinct peaks of the relative horizontal components of Coulomb's force f_h: two of them are at approximately $\pm 55°$ and one is at $0°$.

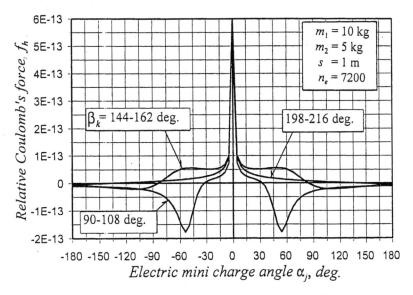

Fig. I5.2 Relative horizontal components of Coulomb's force f_h between electric mini charges representing two gravitational spiral fields.

16. Gravitational Forces Between Two Adjacent Bodies

Let us consider now two adjacent bodies 1 and 2 with respective gravitational masses m_1 and m_2. Fig. 16.1 shows the distribution of electric mini charges related to the gravitational spiral fields of these bodies, and also the lines of electric forces between some electric mini charges.

In relation to body 1, positive electric mini charges e_a representing the inner gravitational spiral field are distributed along a circle with relative spiral radius b_a, while negative electric mini charges e_A representing the outer gravitational spiral field are distributed along a circle with relative spiral radius b_A. In relation to body 2, positive electric mini charges e_b representing the inner gravitational spiral field are distributed along a circle with relative spiral radius b_b, while negative electric mini charges e_B representing the outer gravitational spiral field are distributed along a circle with relative spiral radius b_B. Importantly, in each body, the total electric charge of all electric mini charges is equal to zero. Thus, the gravitational fields of both bodies are electrically neutral.

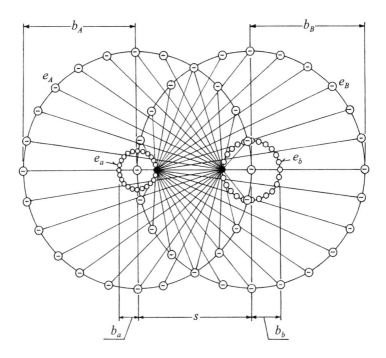

Fig. 16.1 Lines of electric forces between electric mini charges related to gravitational spiral fields of two adjacent bodies.

Parameters of the gravitational fields were determined from Eqs. (I5.1)-(I.5-3). Gravitational electromass ratio Ω was calculated from Eq. (I2-1).

To calculate the relative gravitational force f_g, it is necessary to sum up the relative horizontal components of Coulomb's forces f_h between all the electric mini charges related to the respective gravitational spiral fields. Once the relative gravitational force f_g is calculated, the absolute value of the gravitational force F_g can be obtained from the equation:

$$F_g = f_g \frac{ke_o^2}{r_i^2} \qquad (I6\text{-}1)$$

As an example, Fig. I6.2 shows a distribution of the relative gravitational force f_g between two bodies as a function of the angular position of the electric mini charges α_j. In this case, we considered the bodies with gravitational masses $m_1 = 10$ kg and $m_2 = 5$ kg separated by the distance $S = 1$m.

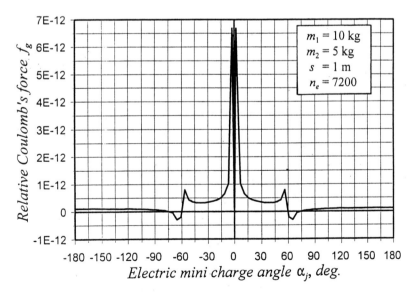

Fig. I6.2 Relative gravitational force f_g between electric mini charges representing two gravitational spiral fields .

Fig. I6.3 shows the results of calculations of the gravitational force F_g between two bodies as a function of distance S between these bodies. We considered three sets of bodies. In the first set, the gravitational masses were: $m_1 = 5$ kg and $m_2 = 5$ kg. In the second set, the gravitational masses were: $m_1 = 10$ kg and $m_2 = 5$ kg. In the third set, the gravitational masses were: $m_1 = 10$ kg and $m_2 = 10$ kg.

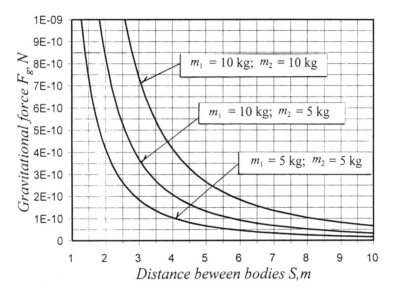

Fig. I6.3 Gravitational force F_g between electric mini charges representing gravitational spiral fields of two adjacent bodies.

The calculation show that the gravitational force F_g calculated based on Coulomb's forces acting between electrical mini charges is exactly the same as the gravitational force calculated based on Newton's inverse square law described by Eq. (I1-1).

REFERENCES

1. Agassi, J., *Faraday as a Natural Philosopher*, The University of Chicago Press, Chicago, ILL, 1971.
2. Davies, P. C. W. and Brown, J., *Superstrings - A Theory of Everything?*, Cambridge University Press, Cambridge, UK, 1988.
3. Greene, B., *The Elegant Universe - Superstrings, Hidden Dimensions,*

and the Quest for the Ultimate Theory, W.W. Norton & Company, New York, 1999.

4. Kaku, M., *Introduction to Superstrings*, Springer-Verlag, New York, 1988.

5. Peat, F. D., *Superstrings and the Search for The Theory of Everything*, Contemporary Books, Lincolnwood , Chicago, IL, 1988.

6. Bartusiak, M., "Gravity Wave Sky," *Discover*, July 1993.

7. Fox, H., "Gravity Waves & Torsion Fields: Faster Than Light?" *Journal of New Energy*, Vol. 3, No. 2/3, 1998.

8. Horgan, J., "Gravity Quantized? - A Radical Theory of Gravity Weaves Space From Tiny Loops," *Scientific American*, September 1992.

9. Rudnicki, K., *Gravitation, Electromagnetism and Cosmology - Toward a New Synthesis*, Apeiron, Montreal, Canada, 2001.

10. Salem, K.G., *The New Gravity - A New Force - A New Mass - A New Acceleration - Unifying Gravity with Light*, Salem Books, Johnstown, PA, 1994.

11. Klyushin, J.G., "On the Maxwell Approach to Gravity," *Report in Seminar of St. Petersburg Society of Physicists*, St. Petersburg, Russia, 1995.

12. Ginzburg, V. B., *Unified Spiral Field and Matter - A Story of a Great Discovery*, Helicola Press, Pittsburgh, PA, 1999.

13. Ginzburg, V. B., *Unified Spiral Nature of the Quantum and Relativistic Universe*, Helicola Press, Pittsburgh, PA, 2002.

J

WHAT'S NEXT?

Contemplating the Next Step

After going so far, you have really opened up a number of choices for yourself regarding what you might do next. Among them are the five possible choices in the order of increasing complexity and excitement:

1. Quit and forget about the idea of a spiral world altogether
2. Quit, but think about this idea at a leisurely pace
3. Read the second part of this book outlining the background of the SFT that requires high-school math
4. Find the gaps in the SFT and work out new solutions
5. Develop your own model of a physical phenomena that is the closest to your heart.

One may find a logical applications of the SFT, for instance, to give a new interpretation of the quantum numbers of atomic electrons, to speculate on various types of stars, and also to explain the radii of planetary orbits, as will be discussed below.

Quantum Numbers of Particles

Quantum numbers of particles are commonly used to define the structure of nuclei, atoms and to explain atomic spectra. These numbers are:

- Principal quantum number n
- Orbital quantum number m_l
- Spin magnetic quantum number m_s.

Spiral field theory also employs three types of quantum numbers for the particles. According to this theory, the principal quantum number is equal to the torix universal excitation level n. The orbital quantum number m_l is determined by the angular orientation of the torices that form the atomic electron. The spin magnetic quantum number m_s is determined by the spin of the torices that form the atomic electron. Physical meaning of the principal, and orbital quantum numbers in the spiral field theory is the same as in the conventional physics. The physical meaning of the spin magnetic quantum number is, however, different.

To understand the difference, let us recollect that the torix is a particular case of helicola. It means that generally it comprises infinite numbers of spirals, one winding around another. In the previous sections of this book, for the sake of simplicity, we limited the number of spirals in the torix to two, one leading spiral in the form a circle and one trailing toroidal spiral.

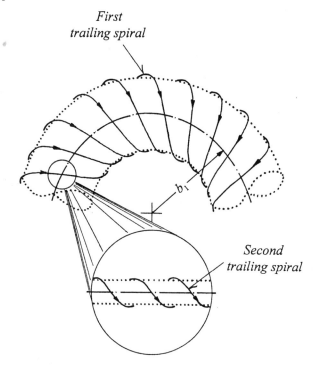

Fig. J1 Torix spin as a direction of vorticity of the second trailing spiral.

However, in application to the fine structure of the particles, we shall consider more than one trailing spiral in the torices. The direction of vorticity of the second trailing spiral defines the *torix spin*, as shown in Fig.

J1. By considering the directions of vorticity of the third and any subsequent higher level of trailing spirals, it may become possible to explain many minute properties of the complex particles and the atomic spectra.

One of the important consequences of the new definition of spin relates to the maximum number of particles that can be in the same quantum state. As defined by Pauli's exclusion principles, two particles with the same principal quantum number n and the same orbital quantum number m_l can be in the same quantum state only if they have opposite spin magnetic quantum number m_s. Since, according to the standard model, one particle has only one value of m_s, the exclusion principal would allow in that case only two particle with opposite values of m_s to be in the same quantum level. This principle explains, for instance, why only two electrons are allowed on the ground atomic quantum level. Also, why only two neutrons are allowed to be in the nucleus of helium He4 (Fig. H4.1).

Since the SFT applies the definition of spin to the torices, the total spin magnetic quantum number m_s of a particle depends on the spin magnetic quantum numbers of its torices. As an example, let us consider only the first levels of the torix spins that comprise an electron, proton shell and neutron shell. Since the electron is made of only one set of real torices (Table F1), it will have only one set of spin magnetic quantum numbers $\pm m_s$. Because the proton shell is also made only one set of real torices (Table F11), it also will have only one set of spin magnetic quantum numbers $\pm m_s$. The neutron shell, however, is made of two sets of real torices (Table F10). Therefore, it must have two sets of spin magnetic quantum numbers $\pm m_{z1}$ and $\pm m_{z2}$. Consequently, four neutrons must be allowed in the same quantum state as shown in Table J1.

Table J1 Allowed spin magnetic quantum numbers of the neutron.

Neutron number	Spin magnetic quantum numbers		
	Outer torix	Inner torix	Neutron
1	$+ m_{z1}$	$+ m_{z2}$	$(m_{z1} + m_{z2})$
2	$- m_{z1}$	$- m_{z2}$	$- (m_{z1} + m_{z2})$
3	$+ m_{z1}$	$- m_{z2}$	$(m_{z1} - m_{z2})$
4	$- m_{z1}$	$+ m_{z1}$	$(- m_{z1} + m_{z2})$

Thus, the SFT explains a possibility for creation of the *nucleon clusters*, and particularly the tetraneutrons that were detected in recent experiments[1].

Real, Inverted, and Imaginary Stars

Earlier, we considered a case that is applicable to real (or visible) stars with outer radii r_b greater than the inversion radii r_j of their gravitational spiral fields. According to the SFT, there are two other types of stars, those with the outer radii r_b smaller than the inversion radii r_j of gravitational spiral fields.

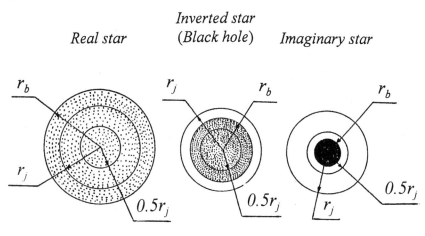

Fig. J2 Three groups of stars.

Thus, all the stars can be divided into three distinct groups (Fig. J2):

Real stars $(r_b > r_j)$ - These are stars like the Sun. In these stars, a sphere with the radius r_i is hidden from the outside world. Therefore, according to Eq. (J2), since $b_1 < 1$, nothing can orbit these stars with the ultimate spiral field velocity C.

Inverted stars $(0.5\,r_j \leq r_b < r_j)$ - These are very dense stars. They are located inside an inverted toroidal spiral field that corresponds to *anti-gravity*. Because in these stars a sphere with the inversion radius r_i is exposed to the outside world, they can be surrounded by bodies moving with the ultimate spiral field velocity C. As a result of anti-gravity, some of these bodies may even be captured by the inverted stars (commonly referred to as *black holes*).

Imaginary stars $(r_b < 0.5\,r_j)$ - These are super-dense stars. The double toroidal spiral field associated with these stars is imaginary. This means that some of its spiral field parameters, including translational velocity, are expressed with imaginary numbers. Importantly, the radius of these stars r_b may be negative. These are the *inverted imaginary stars* in which matter is inverted inside out. They may be among the candidates for *dark matter.*

Quantum Levels of Planets

Titius was probably the first to point out a relationship describing the approximate distances of the planets from the Sun. It is now known as Bode's law[2]. If we write a series of 4's and add to them the numbers 0, 3, $3 \times 2 = 6$, $6 \times 2 = 12$, $12 \times 2 = 24$, etc., we obtain a series of numbers which are approximately ten times the mean distances of the planets from the Sun in astronomical units as shown in Table J1. Bode's law fails for Neptune.

Table J1 Calculation of planetary orbits according to Bode's law.

Mercury	Venus	Earth	Mars	Asteroids	Jupiter	Saturn	Uranus
4	4	4	4	4	4	4	4
0	3	6	12	24	48	96	192
4	7	10	16	28	52	100	196

Alternatively, one can define quantum levels of planets with the method that the SFT employs for atomic electrons. According to the SFT, the double-toroidal spiral fields that form atomic electrons have the same inversion radius r_i that is defined by Eq. (H1-5) of Section H. Similarly, in the planetary system, the gravitational spiral fields associated with all the planets of this system have the same inversion radius r_j that is defined by Eq. (I3-2) of Section I.

You may want to investigate if the quantum values of the relative spiral radii b_1 of both electrons and planets are expressed by the same relationship as the function of universal excitation level n and universal spiral field constant U:

$$b_1 = 2(1 + n^2 U^2) \qquad n = 1, 2, 3, \ldots \qquad (J1)$$

The only difference between quantum levels of electrons and planets is in the magnitudes of universal excitation level n. For instance, in the hydrogen atom, the ground level of an electron corresponds to universal excitation level $n = 1$, and all the subsequent quantum energy levels of the magnitudes of n will be equal to 2, 3, 4, etc.

Table J2 Quantum levels n of planets of the solar system ($r_i = 738.38$ m).

Planets	Mean distance to Sun[3], m	b_1	n	Distance error, %
Mercury	5.790920×10^{10}	7.8427×10^{07}	46	-1.26
Venus	1.082089×10^{11}	1.4655×10^{08}	62	1.56
Earth	1.495979×10^{11}	2.0260×10^{08}	73	1.28
Mars	2.279410×10^{11}	3.0870×10^{08}	90	1.53
Asteroids	4.328358×10^{11}	5.8620×10^{08}	125	-0.06
Jupiter	7.783284×10^{11}	1.0541×10^{09}	167	0.69
Saturn	1.426990×10^{12}	1.9326×10^{09}	225	1.69
Uranus	2.869572×10^{12}	3.8863×10^{09}	320	1.10
Neptune	4.496584×10^{12}	6.0898×10^{09}	400	1.40
Pluto	5.900141×10^{12}	7.9907×10^{09}	460	0.60

Table J3 Quantum levels n of some satellites of Jupiter ($r_i = 0.705205$ m).

Satellites	Mean distance to Jupiter[3], m	b_1	n	Distance error, %
Io	4.216×10^{08}	5.9874×10^{08}	125	1.93
Europa	6.709×10^{08}	9.5135×10^{08}	160	-1.00
Ganymede	1.070×10^{09}	1.5173×10^{08}	200	1.05
Gallisto	1.880×10^{09}	2.6659×10^{08}	267	-0.38

Table J4 Quantum levels n of some satellites of Uranus ($r_i = 0.032255$ m).

Satellites	Mean distance to Uranus[3], m	b_1	n	Distance error, %
Miranda	1.294×10^{08}	1.8349×10^{08}	70	-0.24
Ariel	1.910×10^{08}	2.7084×10^{08}	85	-0.14
Umbriel	2.663×10^{08}	3.7762×10^{08}	100	0.60
Titania	4.359×10^{08}	6.1812×10^{08}	130	-2.57
Oberon	5.835×10^{08}	8.2742×10^{08}	150	-2.03

As an example, Tables J2 - J4 show quantum levels n for the planets of the solar system and for some satellites of Jupiter and Uranus, and the errors of the calculated distances based on the SFT.

Unlike the hydrogen atom, the magnitudes of the quantum levels n of the planetary and satellite systems are much greater. For instance, in the solar system, their range is from 46 for the planet Mercury, which is nearest to the Sun, to 460 for the most distant planet Pluto, as shown in Table J2.

A New Definition of Age

How old are we? What is the age of the things that surround us? According to the conventional definition, the age of an entity is equal to the calendar time elapsed since the inception of the entity. For instance, if a person was born on January 1, 1950, then in January 1, 2000 he was fifty years old. Why fifty? Because during this period of time, the Earth made fifty revolutions around the Sun. The same definition is applied to the age of other entities. For instance, if an electron was created one billion calendar years ago then since its creation the Earth had revolved one billion times around the Sun.

Thus, according to the conventional definition, the age of the entities has nothing to do with the internal processes that take place inside these entities. Simply count the revolutions of the Earth around the Sun and you will determine their age. Intuitively, we feel that there is something wrong with this definition of age. We commonly relate the age of people as well as other things that surround us to their appearances. Older people look certainly different than young ones, and used cars look much less attractive than brand new ones. But what about an electron? Unfortunately, here we fail to associate the age of an electron with its appearance. Physicists tell us that today an electron looks exactly the same as it did right after its creation. But the electron is not unique. The nucleons, proton and neutron, that form atomic nuclei also stay unchanged over time. So, why does everything change over time but these three elementary particles? We can explain this fact by introducing a new definition of age, with some help from the SFT and other theories[4-13].

As we discussed in the previous sections of this book, the spiral energy of the torices that make up the electron, proton and neutron can be either positive and negative. In each of these particles, negative energy absorbed by imaginary spiral fields is equal to positive energy "spent" by real spiral fields. As an example, Fig. J3 shows one cycle of the "inhaling and exhaling process" of an electron. A solid line depicts electron energy, and a dotted line depicts electron *internal age* that is proportional to electron energy. In the beginning of each cycle, the electron energy and its internal age are equal to zero. During the first part of the cycle, when electron absorbs negative energy, the electron initially becomes younger and then, after reaching a maximum negative value, its internal age reduces back to zero.

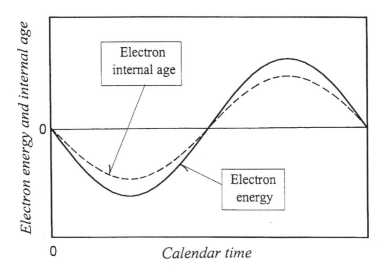

Fig. J3. One cycle of the electron energy and internal age versus calendar time.

During the second part of the cycle, when electron spends positive energy, the electron initially becomes older and then, after reaching a maximum positive value, its internal age reduces back to zero. Since within each cycle electron internal age oscillates around zero, it never grows older internally as calendar time goes on. For the same reason, protons and neutrons (within nuclei) are also ageless. Notably, the cycle times of these particles are extremely short, on the order of 10^{-18} seconds for electrons, and on the order of 10^{-23} seconds for protons and neutrons

Unfortunately, we are not as lucky as electrons, protons and neutrons. We do become older, but still our internal clock works in a completely different way than the conventional clock based on calendar time. Amazingly, according to our internal clock, we become younger before we grow older. This becomes clear if we consider positive and negative components of energy on different structural levels of a human body. As we established earlier, on atomic level, positive and negative energies are completely balanced. Thus, three principal particles that make up all the atoms contained in our bodies are internally ageless. However, total energy of a human body depends not only on the balance of energy on atomic level, but also on the balance of biochemical energy related to molecular, cellular and other structural levels of a human body.

A good example of the biochemical energy is the chemical energy that is absorbed in the form of oxygen and released in the form of carbon dioxide during breathing. The cycle time of this process is counted in seconds, making it about 10^{18} longer than cycle time of electrons. The cycle time for biochemical energy responsible for processing of food is counted in hours. Another kind of biochemical energy is absorbed and released by our brain cells during learning, memorizing, thinking and controlling bodily functions. In this case, the cycle time may range from milliseconds to hours, days, or even years.

A sum of all the periodically oscillating components of energy of a human body produces the *life cycle of a human body*. As shown in Fig. J4, in the beginning of human life, the total energy of a human body is equal to zero. Also equal to zero is its internal age. As the human body starts to grow, it initially absorbs more biochemical energy than it spends. Consequently, the total energy of the human body stays negative, making this body younger. Probably, at the age of its physical, mental and sexual maturity, the development of the human body reaches a point when it spends as much biochemical energy as it absorbs. At this time, both the total body energy and its internal age become equal to zero. Beginning

What's Next?

from this time, the total body energy stays positive, making this body older. At the end of the cycle the aging process slows down, and eventually the body total energy and its internal age return back to zero.

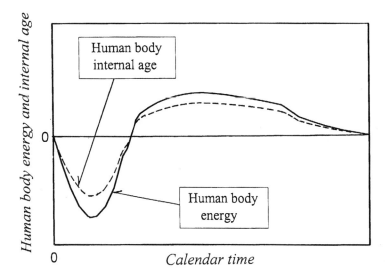

Fig. J4. Life cycle of a human body.

There are three important consequences of the new interpretation of age in respect to human beings. Firstly, this interpretation confirms the law of conservation of energy and time: the human life starts and ends at zero energy level and at zero internal age. The second important consequence arises because both the energy of a human body and its internal age are very close to zero level three times during its life time: (1) shortly after conception, (2) during transition of a human being from adolescence to adult life, and (3) at the end of life cycle. In all three cases, a human being behaves in a similar way, "like a child." This explains why during these three periods of life he or she must be treated accordingly.

The second important consequence is related to the fact that the internal age of a human body does not coincide with its calendar age. Moreover, internal ages are different for people having the same calendar age. Those who are able to consume more negative biochemical energy during early period of their lives have greater opportunities of either spending this energy in more productive way or to live longer, or both. A good news is that we can change the balance of energy inside our bodies during any period of our life, striving to increase influx of negative energy while reducing a wasteful part of positive energy. This can be achieved by further optimization of our life style, leisure habits, diet, environment, working attitude, and positive state of mind. In short:

We have significant control over our internal time clock.

Create Your Own Theory and Philosophy

The author will certainly be pleased if you will apply the SFT to investigate some of the topics of your own interest. He, however, will be pleased just as well if you decide to develop your own theory. The last choice is the most difficult one, but also the most rewarding. It does not matter whether you are a professional scientist or an amateur of science. You are entitled to have your own theory about how our universe works. Do not believe anyone who will tell you otherwise.

Spiral field theory and its concept of helicola helps us to think cosmically, and to open our mind for asking various daring philosophical questions. Here is the one that may puzzle you: "What is the purpose of the human race in the development of the universe?" Are we simply here for a ride, or we are destined to play a very important and still unimaginable role that will affect the future of the entire universe? In Appendix *B* of this book, I outlined one of many possible answers to this question. You may propose your own one. There is a place at the end of this Section where you can write your own ideas and comments.

REFERENCES

1. *New Scientist,* vol. 176, Issue 2336, October 2002, p. 30.
2. Duncan, J. C., *Astronomy - A Textbook*, Fifth Edition, Harper & Brothers Publishers, New York, 1926.
3. Zombeck, M. V., *Handbook of Space Astronomy and Astrophysics*, Second Edition, Cambridge University Press, Cambridge, UK, 1990.
4. Riordan, M. and Schramm, D.N., *The Shadows of Creation - Dark Matter and the Structure of the Universe*, W.H. Freeman and Company, New York, 1991.
5. Van Flandern, T., *Dark Matter, Missing Planets & New Comets- Paradoxes Resolved, Origins Illuminated*, North Atlantic Books,

Berkeley, CA, 1993.

6. Krauss, L., *Quintessence - The Mystery of Missing Mass in the Universe*, Basic Books, New York, NY, 2000.

7. Boslough, J., *Masters of Time - Cosmology at the End of Innocence*, Addison-Wesley Publishing Company, Reading, MA, 1992.

8. Hawking, S., *Black Holes and Baby Universes and Other Essays*, Bantam Books, New York, 1993.

9. Novikov, I. D., *The River of Time*, Cambridge University Press, Cambridge, UK, 1998.

10. Price, H., *Time's Arrow and Archimedes' Point*, Oxford University Press, New York, 1996.

11. Russell, P., *The White Hole in Time*, Harper San Francisco, 1992.

12. Thorne, K. S., *Black Holes & Time Warps - Einstein's Outrageous Legacy*, W. W. Norton & Company, New York, 1994.

13. *Scientific American*, Sept. 2002, pp. 36-93.

Your own ideas and comments

Please feel free to write below your own ideas and comments.

LIST OF PUBLISHED PAPERS CONSULTED

A

Abramowicz, M.A., "Black Holes and the Centrifugal Force Paradox," *Scientific American*, March 1993.

Adleman, L.M., "Computing with DNA," *Scientific American*, August 1998.

Adler, C.G., "Does Mass Really Depend on Velocity, Dad?," *American Journal of Physics* 55(8), August 1987.

Aftergood, S., et al, "Nuclear Power in Space," *Scientific American*, June 1991.

Akimov, A.E. and Shipov, G.I., "Torsion Fields and Their Experimental Manifestations," *Proc. of the Int'l Conference on New Ideas in Natural Sciences*, St. Petersburg, Russia, June 1996.

Akimov, A.E. and Tarasenko, V.Y., "Models of Polarized States of the Physical Vacuum and Torsion Fields," *Fizika*, No.3, March 1992.

Albert, D.Z., "Bohm's Alternative to Quantum Mechanics," *Scientific American*, May 1994.

Alfven, H., "Model of Plasma Universe," *IEEE Transactions on Plasma Science*, Vol. PS-14, No. 6, June 1986.

Ariza, L.M., "Einstein's Drag," *Scientific American*, July 1998.

Awschalom, D.D., et al, "Spintronics," *Scientific American*, June 2002.

B

Baeyer, H.C., "The Cinderella Particle", *Discover*, December 1993.

Baeyer, H.C., "Black Holes Ants," *Discover*, July 1995.

Barton, G. and Dombey, N., "Casimir Effect for Massive Photons," *Nature*, Vol. 311 (27), Sept. 1984.

Bartusiak, M., "If You Like Black Holes, You'll Love Cosmic Strings," *Discover*, April 1988.

Bartusiak, M., "Loops of Space," *Discover*, April 1993.

Bartusiak, M., "Gravity Wave Sky," *Discover*, July 1993.

Bearden, T.E., "Dark Matter or Dark Energy?" *Journal of New Energy*,

Vol. 4, No. 4, 2000.

Beller, M., "The Sokal Hoax: At Whom Are We Laughing?" *Physics Today*, September 1998.

Bergman, D.L. and Wesley, J.P., "Spinning Charged Ring Model of Electron Yielding Anomalous Magnetic Moment," *Galilean Electrodynamics*, Vol.1, No.5, Sept./Oct.1990.

Bergman, D.L., "Spinning Charged Ring Model of Elementary Particles," *Galilean Electrodynamics*, Vol. 2, No. 2, March/April 1991.

Bergman, D.L., "New Spinning Charged Ring Model of the Electron," *Twin-Cities Creation Conference*, July 29 - August 1, 1992.

Bergman, D.L., "The Stable Elementary Particles," *Twin-Cities Creation Conference*, July 29 - August 1, 1992.

Bergman, D.L., "Commentary on Sub-Quantum Physics," *Galilean Electrodynamics*, Vol. 8, No. 5, Sept./Oct.1997.

Bergman, D.L., "Conflict of Atomism and Creationism in History," *IVth Int. Conference on Creationism*, Pittsburgh, PA, 1998.

Bergman, D.L., "Comparison of Physical Models and Electromagnetic Field Theory to Quantum Mechanics and Theories of Relativity," *Proc. of Physical Interpretations of Relativity Theory*, London, Sept. 1998.

Bertozzi, W., "Speed and Kinetic Energy of Relativistic Electrons," *American Journal of Physics* Vol. 32, No.7, July 1964.

Boden, M.A., "Artificial Genius," *Discover*, October 1996.

Boivin, J., "The Heart Single Field Theory," *Speculations in Science and Technology*, Vol. 3, No. 2, 1980.

Bostick, W.H., "Experimental Study of Ionized Matter Projected Across a Magnetic Field," *Physical Review*, Vol. 104, No. 2, Oct.1956.

Bostick, W.H., "Experimental Study of Plasmoids," *Physical Review*, Vol. 106, No. 3, May 1957.

Bostick, W.H., "Plasmoids," *Scientific American*, Vol. 197, No. 4, Oct. 1957.

Bostick, W.H., "Possible Hydromagnetic Simulation of Cosmical Phenomena in the Laboratory," *Review of Modern Physics*, Vol. 30, No. 3, July 1958.

Bostick, W.H., et al, "Pair Production of Plasma Vortices," *Physics of Fluids*, Vol. 9, No.10, October 1966.

Bostick, W.H., "The Morphology of the Electron," *International Journal of Fusion Energy*, Vol. 3, No. 1, January 1985.

Bostick, W.H., "What Laboratory-Produced Plasma Structures Can Contribute to the Understanding of Cosmic Structures Both Large and Small,"

IEEE Transactions on Plasma Science, Vol. PS-14, No.6, December 1986.

Bostick, W.H., "Stockholm, August 1956, Revisited," *IEEE Transactions on Plasma Science*, Vol. 17, No.2, April 1989.

Bostick, W.H., "Mass, Charge, and Current: The Essence and Morphology," *Physics Essays*, Vol. 4, No. 1, 1991.

Boyer, T.H., "Electrostatic Potential Energy Leading to an Inertial Mass Change for a System of Two Point Charges," *Am. J. Phys.* 46 (4), Apr. 1978.

Brill, D.R. and Wheeler, J.A., "Interaction of Neutrinos and Gravitational Fields," *Reviews Of Modern Physics*, Vol. 29, No. 3, July 1957.

Brown, L. and Gabrielse, G., "Geonium Theory: Physics of a Single Electron or Ion in a Penning Trap," *Reviews Of Modern Physics*, Vol. 58, No. 1, January 1986.

Bush, V., "The Force Between Moving Charges," *Journal of Mathematics and Physics* (MIT), Vol. 5, 1925-26 .

C

Casten, R.F. and Feng, D.H., "Nuclear Dynamical Supersymmetry," *Physics Today*, November 1984.

Chambers, L.G., "The Hund Gravitational Equations and Geons," *Can. J. Phys.* Vol. 37, 1959.

Ciufolini, I., et al, "Test of General Relativity and Measurement of the Lense - Thirring Effect with Two Earth Satellites," *Science*, Vol. 279, March 1998.

Cole, K.C., "Escape From 3-D: Visiting Higher Dimensions," *Discover*, July 1993.

Collins, G.P., "Supersymmetric QCD Sheds Light On Quark Confinement and the Topology of 4 - Manifolds," *Physics Today*, March 1995.

Cornell, E.A., et al, "Single-Ion Cyclotron Resonance Measurement of $M(CO^+)/M(N_2^+)$," *Physical Review Letters*, Vol. 63, No. 16, October 1989.

D

D'Agnese, J., "The Last Big Bang Man Left Standing," *Discover*, July 1999.

Diamond, R. M. and Stephens, F. S., "Angular Velocity: A New Dimension In Nuclei," *Nature*, Vol. 310 (9), August 1984.

Deutsch, D. and Lockwood, M., "The Quantum Physics of Time Travel," *Scientific American*, March 1994.

Dingle, H., "The Case Against Special Relativity," *Nature,* Vol. 216, October 1967.

Disney, M., "A New Look at Quasars," *Scientific American,* June 1998.

Dmitriyev, V.P., "Mechanical Analogy for the Wave - Particle: Helix on Vortex Filament," *Apeiron,* Vol. 8, No. 2, April 2001.

Drew, H.R., "The Electron as a Four-Dimensional Helix of Spin - 1/2 Symmetry," *Physics Essays,* Vol. 12, No. 4, 1999.

E

Ekstrom, P. and Wineland, D., "The Isolated Electron," *Scientific American,* Vol. 243, August 1980.

Elderfield, J., "The Line of Free Men: Tatlin's 'Towers' and the Age of Invention," *Studio International,* November 1969, Vol. 178, No. 916.

Elmer-Dewitt, P., "The Genetic Revolution," *Time,* January 1994.

Enzer, D. and Gabrielse, G., "Dressed Coherent States of the Anharmonic Oscillator," *Physical Review Letters,* Vol. 78, No. 7, February 1997.

Ericsson, J., *Propelling Steam Vessels,* U.S. Patent No. 588, February 1, 1838.

Ernst, F.J., "Linear and Toroidal Geons," *Physical Review,* Vol. 105, No. 5, March 1957.

Ernst, F.J., "Variational Calculations in Geon Theory," *Physical Review,* Vol. 105, No. 5, March 1957.

F

Faber, S., "Sky Rivers," *Discover,* January 1994.

Falthammar, C.G., "Plasma Physics from Laboratory to Cosmos - The Life and Achievements of Hannes Alfven," *IEEE Trasactions on Plasma Science,* Vol. 25, No. 3, June 1997.

Farley, F.J.M., "Is The Special Theory Right or Wrong?," *Nature,* Vol. 217, January 1968.

Finkler, P., "Relativistic Momentum," *Am. J. Phys.* 64 (15), May 1996.

Fox, H., "Gravity Waves & Torsion Fields: Faster Than Light?" *Journal of New Energy,* Vol. 3, No. 2/3, 1998.

Fox, H. and Jin, S.X., "Low-Energy Nuclear Reactions and High-Density Charge Clusters," *Journal of New Energy,* Vol. 3, No. 2/3, 1998.

Freedman, D.H., "Cosmic Time Travel," *Discover,* June 1989.

Freedman, D.H., "The Mysterious Middle of the Milky Way," *Discover,* November 1998.

G

Gabrielse, G., et al, "Observation of a Relativistic, Bistable Hysteresis in the Cyclotron Motion of a Single Electron," *Physical Review Letters*, Vol. 54, No. 6, February 1985.

Gabrielse, G., et al, "Thousandfold Improvement in the Measured Antiproton Mass," *Physical Review Letters*, Vol. 65, No. 11, Sept. 1990.

Gabrielse, G., "Extremely Cold Antiprotons," *Scientific American*, Vol. 267, December 1992.

Gabrielse, G., et al, "Special Relativity and the Single Antiproton: Forty-fold Improved Comparison of Antiproton and Proton Charge-To-Mass Ratios," *Physical Review Letters*, Vol. 74, No. 18, May 1995.

Gabrielse, G., "Relativistic Mass Increase at Slow Speeds," *Am. J. Phys.* 63 (6), June 1995.

Gao, J., et al, "The Ponderomotive Four-Momentum," *Can. J. Phys.* 76, 1998.

Gatti, F., et al, "Detection of Environmental Fine Structure in the Low-energy β-decay Spectrum of [187]Re," *Nature*, Vol. 397, January 1999.

Gingerich, O., "Cosmology +1," *Readings from Scientific American*, March 1988.

Ginsparg, P. and Glashow, S., "Desperately Seeking Superstrings," *Physics Today*, May 1986.

Ginzburg, V.B., *Multiple-Level Universe*, IRMC, Pittsburgh, PA, 1993.

Ginzburg, V.B., "Toroidal Spiral Field Theory," *Speculations in Science and Technology*, Vol. 19, 1996.

Ginzburg, V.B., "Structure of Atoms and Fields," *Speculations in Science and Technology*, Vol. 20, 1997.

Ginzburg, V.B., "Double Helical and Double Toroidal Spiral Fields," *Speculations in Science and Technology*, Vol. 22, 1998.

Ginzburg, V.B., "Nuclear Implosion," *Journal of New Energy*, Vol. 3, No. 4, 1999.

Ginzburg, V.B., *Continuous Spiral Motion System for Rolling Mills*, U.S. Patent No. 5,970,771, Oct. 26, 1999.

Ginzburg, V.B., *Continuous Spiral Motion System and Roll Bending System for Rolling Mills*, U.S. Patent No. 6,029,491, Feb. 29, 2000.

Ginzburg, V.B., "Unified Spiral Field, Matter and Ether - An Introduction to Spiral Field Theory, *Conference Storrs 2000*, University of Connecticut, June 4-9, 2000.

Ginzburg, V.B., "Dynamic Aether," *Journal of New Energy*, Vol. 6, No. 1, 2001.

Ginzburg, V.B., "Electric Nature of Strong Interactions," *Journal of New Energy*, Vol. 7, No. 2, 2002.

Green, M., "Superstrings," *Scientific American*, Vol. 255 (3), Sept. 1986.

Grunstein, M., "Histones as Regulators of Genes," *Scientific American*, October 1992.

Gulko, A.G. and Bergman, D.L., "Charged Ring Model of Elementary Particles: A Controversy," *Galilean Electrodynamics*, Vol. 4, No. 5, Sept./Oct.1993.

H

Haber, H.E. and Kane, G.L., "Is Nature Supersymmetric," *Scientific American*, Vol. 254 (6), June 1986.

Hall, D.S. and Gabrielse, G., "Electron Cooling of Protons in a Nested Penning Trap," *Physical Review Letters*, Vol. 77, No. 10, Sept. 1996.

Hillard, D. and Gray, J., "The Ubiquitously Expanding Universe - An Alternative Formulation to Newtonian Gravity," *Speculations in Science and Technology*, Vol. 17, No. 3, 1994.

Horgan, J., "Gravity Quantized? - A Radical Theory of Gravity Weaves Space From Tiny Loops," *Scientific American*, September 1992.

Horgan, J., "Questioning the 'It From Bit'," *Scientific American*, June 1991.

Horgan, J., "Quantum Philosophy," *Scientific American*, July 1992.

Horgan, J., "The Death of Proof," *Scientific American*, October 1993.

Hotson, D.L., "Dirac's Equation and the Sea of Negative Energy, Part 1," *Infinite Energy*, Vol. 8, Issue 43, 2002.

Hotson, D.L., "Dirac's Equation and the Sea of Negative Energy, Part 2," *Infinite Energy*, Vol. 8, Issue 44, 2002.

I

Icke, V., "From Expansion to Intelligence in the Universe," *Speculations in Science and Technology*, Vol. 14, No. 4, 1991.

J

Jin, S.X, and Fox, H., "Characteristics of High-Density Charge Clusters: A Theoretical Model," *Journal of New Energy*, Vol. 1, No. 4, 1996.

Johnson, P.O., "Ball Lightning and Self-Containing Electromagnetic Fields," *American Journal of Physics*, Vol. 33, 1965.

Joyce, G.F., "Directed Molecular Evolution," *Scientific American*,

December 1992.

K

Kaplan, A.E., "Hysteresis in Cyclotron Resonance Based on Weak Relativistic-Mass Effects of the Electron," *Physical Review Letters*, Vol. 48, No. 3, January 1982.
Kaplan, A.E. and Elici, A., "Hysteretic (bistable) Cyclotron Resonance in Semiconductors," *Physical Review B*, Vol. 29, No. 2, January 1984.
Kaplan, A.E., "Ultimate Bistability: Hysteretic Resonance of a Slightly-Relativistic Electron," *IEEE Journal of Quantum Electronics*, Vol. QE21, No. 9, Sept. 1985.
King, M.B., "Vortex Filaments, Torsional Fields and the Zero-Point Energy," *Journal of New Energy*, Vol. 3, No. 2/3, 1998.
King, M.B., "Dual Vortex Forms: The Key to a Large Zero-Point Energy Coherence," *Journal of New Energy*, Vol. 5, No. 2, 2000.
Klauder, J. and Wheeler, J. A., "On the Question of a Neutrino Analog to Electric Charge," *Review of Modern Physics*, Vol. 29, No. 3, July 1957.
Kobayashi, B., "The Exploring Eye: Spiral Tower," *The Architectural Review*, Vol. 125, No. 748, May 1959.

L

Lamb, G.L., "Solutions and the Motion of Helical Curves," *Physical Review Letters*, Vol. 37, No. 5, August 1976.
Lakhtakia, A., "Viktor Trkal, Beltrami Fields, and Trkalan Flows," *Czechoslovak Journal of Physics*, Vol. 44, No. 2, 1994.
Lakhtakia, A. and Weiglhofer, W.S., "Time-Dependent Beltrami Fields in Free Space: Dyadic Green Functions and Radiation Potentials," *Physical Review E*, Vol. 49, Number 6, June 1994.
Lakhtakia, A. and Weiglhofer, W.S., "Covariances and Invariances of the Beltrami-Maxwell Postulates," *IEE Proc. - Sci. Meas. Technol.*, Vol. 142, No. 3, May 1995.
Lande, A., "Quantum Fact and Fiction," *American Journal of Physics*, Vol. 33, 1965.
Lerner, E.J., "Magnetic Vortex Filaments, Universal Scale Invariants, and the Fundamental Constants," *IEEE Trasactions on Plasma Science*, Vol. PS-14, No. 6, December 1986.
Linde, A., "The Self-Reproducing Inflationary Universe," *Scientific American*, November 1994.

Littmann, C.R., "Particle Mass Ratios and Similar Volumetric Ratios in Geometry," *Journal of Chemical Information and Computer Sciences*, 1995, 35.

Lucas, C.W. and Lucas, J.C., "Weber's Force Law for Relativistic Finite-Size Elastic Particles," *Journal of New Energy*, Vol. 5, No. 3, 2001.

M

Maddox, J., "Alternatives to the Big Bang," *Nature*, Vol. 308, April 1984.

McCrea, W.H., "Why the Special Theory of Relativity is Correct," *Nature*, Vol. 216, October 1967.

McCrea, W.H., "Arthur Stanley Eddington," *Scientific American*, June 1991.

McGoveran, D.O. and Noyes, H.P., "On the Fine-Structure Spectrum of Hydrogen," *Physics Essays*, Vol. 4, No. 1, 1991.

Meyerhofer, D.D., et al, "Observation of Relativistic Mass Shift Effects During High-Intensity Laser-Electron Interactions," *J. Opt. Soc. Am. B*, Vol. 13, No. 1, January 1996.

Mermin, N.D., "What is Quantum Mechanics Trying to Tell Us?," *Am. J. Phys.* 66 (9), September 1998.

Monastersky, R., "Mysteries of the Orient," *Discover*, April 1993.

Mugnai, D., et al, "Observation of Superluminal Behaviors in Wave Propagation," *Physical Review Letters*, Vol. 84, Number 21, May 2000.

N

Nardi, V., "Structure and Propagation of Electron Beams," *Proc. 2nd Int. Conf. Energy Storage, Compression, and Switching*, Venice, Italy, 1978.

Nardi, V., et al, "Internal Structure of Electron-Beam Filaments," *Physical Review A*, Vol. 22, No. 5, November 1980.

Nauenberg, M., et al, "The Classical Limit of an Atom," *Scientific American*, June 1994.

Normille, D., "Japan Readies Helical Device to Probe Steady-State Plasmas," *Science*, Vol. 279, March 1998.

O

Okun, L.B., "The Concept of Mass," *Physics Today*, Vol. 42, June 1989.

Okun, L.B., "The Concept of Mass (mass, energy, relativity)," *Sov. Phys. Usp.* 32 (7), July 1989.

Oliwensrein, L., "Bent out of Shape," *Discover*, July 1993.

Osterbrock, D.E., et al, "Edwin Hubble and The Expanding Universe," *Scientific American*, July 1993.

P

Parker, S., "Relativity in an Undergraduate Laboratory - Measuring the Relativistic Mass Increase," *American Journal of Physics*, Vol. 40, No. 2, February 1992.

Peratt, A.L., "Simulating Spiral Galaxies," *Sky & Telescope*, August 1984.

Peratt, A.L., "Birkeland and the Electromagnetic Cosmology," *Sky & Telescope*, May 1985.

Potemra, T.A. and Peratt, A.L., "Guest Editorial - The Golden Anniversary of 'Magnetic Storms and Aurorae'," *IEEE Trasactions on Plasma Science*, Vol. 17, No. 2, April 1989.

Power, J.A. and Wheeler, J.A., "Thermal Geons," *Reviews of Modern Physics*, Vol. 29, No.3, July 1957.

Puthoff, H.E., "The Energetic Vacuum: Implication for Energy Research," *Speculations in Science and Technology*, Vol. 13(4), 1991.

Puthoff, H.E., et al, "Engineering the Zero-Point Field and Plarizabale Vacuum for Interstellar Flight," *Journal of New Energy*, Vol. 6, No. 1, 2001.

R

Reed, D., "Archetypal Vortex Topology in Nature," *Speculations in Science and Technology*, Vol. 17, No. 3, 1994.

Reed, D., "The Vortex as Topological Archetype - A Key to New Paradigms in Physics and Energy Science," *Proceedings of the 4th Symposium on New Energy*, May 1997.

Reed, D., "Novel Electromagnetic Concepts and Implications for New Physics Paradigms and Energy Technologies," *Journal of New Energy*, Vol. 2, No. 1, Spring 1997.

Reed, D., Excitation and Extraction of Vacuum Energy Via EM-Torsional Field Coupling - Theoretical Model," *Journal of New Energy*, Vol. 3, No. 2/3, 1998.

Rhodes, D. and Klug, A., "Zinc Fingers," *Scientific American*, February 1993.

Robertson, D.S., "Speculation on the Nature of the Atom," *Speculations in Science and Technology*, Vol. 17, No. 1, 1994.

Rohrlich, F., "The Dynamics of a Charged Sphere and the Electron," *Am. J. Phys.* 65 (11), November 1997.

Roman, M., "Tornado Tracker," *Discover*, June 1989.

Rothman, T., "Japanese Temple Geometry," *Scientific American*, May 1998.

S

Sahtouris, E.S., "The Consciousness of Living Nature," *The Cosmic Light*, Vol. 4, No. 2, Spring 2002.

Sallhofer, H., "Hydrogen in Electrodynamics. I.. Preliminary Theories," *Verlag der Zeitschrift fur Naturforschung*, Vol. 43a, December 1988.

San, M.G., "An 'Explicit' Solution of Force-Free Magnetic Fields," *Speculations in Science and Technology*, Vol. 17, No. 2, 1994.

Sandin, T.R., "In Defense of Relativistic Mass," *Am. J. Phys.* 59 (11), November 1991.

Sarg, S., "Abstract of Thesis: Basic Structures of Matter," *BSM Abstract Paper*, October 2001.

Schwattschneider, D., "Escher's Metaphors," *Scientific American*, November 1994.

Schwarzschield, B., "Antiprotons Cooled to 4K and Weighed in a Penning Trap," *Physics Today*, July 1990.

Schwinger, J., "Electromagnetic Mass Revisited," *Foundations of Physics*, Vol. 13, No. 3, 1983.

Scully, M.O. and Sargent, M., "The Concept of the Photon," *Physics Today*, March 1972.

Sen, A., "How Galileo Could Have Derived the Special Theory of Relativity," *Am. J. Phys.* 62 (2), February 1994.

Sharon, N. and Lis, H., "Carbohydrates in Cell Recognition," *Scientific American*, January 1993.

Sheffield, J., "The Physics of Magnetic Fusion Reactors," *Reviews of Modern Physics*, Vol. 66, No.3, July 1994.

Shoulders, K.R., "Energy Conversion Using High Charge Density," *U.S. Patent* No. 5,018,180, May 1991.

Singal, A.K., "The Equivalence Principle and an Electric Charge in a Gravitational Field. II. A Uniformly Accelerated Charge Does Not Radiate," *General Relativity and Gravitation*, Vol..29, No.11, 1997.

Steinberg, M.S., "Charge Invariance and Electron Mass From the Ultimate Speed Experiment," *American Journal of Physics*, Vol. 40, No. 2, February 1992.

Sternglass, E.J., "Evidence for a Possible Common Origin of Baryonic and Non-Baryonic Dark Matter," *Proc. of the 5th Ann. Astrophysics Conference* *in Maryland*, October 1994.

Stewart, I., "Daisy, Daisy, Give Me Your Answer, Do," *Scientific American*, January 1995.

Stoeffl, W., "A New Twist in β-decay," *Nature*, Vol. 397, January 1999.

Strogartz, S. and Stewart, I., "Coupled Oscillators and Biological Synchronization," *Scientific American*, December 1993.

Sugiyama, T., et al, "New 100-in. OD Spiral Pipe Mill at Fukuyama Works," *Iron and Steel Engineer Year Book*, 1978.

Susskind, L., "Black Holes and the Information Paradox," *Scientific American*, April 1997.

Synge, J.L., "The Electrodynamic Double Helix," In *Magic Without Magic: John Archibald Wheeler* - A Collection of Essays in Honor of His Sixtieth Birthday, edited by John R. Klauder, W. H. and Company, San Francisco, 1972.

T

Tarle, G. and Swordy, S. P., "Cosmic Antimatter," *Scientific American*, April 1998.

Taubes, G., "The Anti-Matter Mission," *Discover*, April 1996.

Taubes, G., "Echo of the Big Bang," *Discover*, November 1997.

Tewari, P., Creation of Galactic Matter, and Dynamics of Cosmic Bodies Through Spatial Velocity-Field," *Journal of New Energy*, Vol. 3, No. 4, 1999.

Tilton, H.B., "A Neoclassical Derivation of the Relativistic Factor," *Speculations in Science and Technology*, Vol. 16, No. 4, 1994.

Tsai, L., "The Relation Between Gravitational Mass, Inertial Mass, and Velocity," *Am. J. Phys.* 54 (4), April 1986.

Tseng, C.H. and Gabrielse, G., "One-Electron Parametric Oscillator," *Applied Physics B 60*, 1995.

V

Valone, T., "Inside Zero Point Energy," *Journal of New Energy*, Vol. 5, No. 4, 2001.

Vandyck, M.A., "Putting to Rest Mass Misconceptions," *Physics Today*, Vol. 43, May 1990.

Vestergaard, B., and Javanainen, J., "Semiclassical Analysis of the Metastable Driven and Damped Quantized Anharmonic Oscillator," *Physical Review A*, Vol. 58, No. 2, August 1998.

Vogel, S., "Life in a Whirl," *Discover*, August 1993.

W

Wagner, O.E., "Structure in the Vacuum," *Frontier Perspectives*, Vol. 10, No. 2, Fall 2001.

Walters, J.P., "Spark Discharge: Application to Multielement Spectrochemical Analysis," *Science*, Vol. 198, No. 4319, November 1977.

Ward, P.D., "Coils of Time," *Discover*, March 1998.

Wasserman, J., "On the Structure of Electron." *Speculations in Science and Technology*, Vol. 15, No. 3, 1992.

Waters, T., "The Unfolding World of Chuck Hoberman," *Discover*, March 1992.

Wheeler, J.A., "Geons," *Physical Review*, Vol. 97, No. 2, January 1955.

Wheeler, J. A., "Geometrodynamics and the Problem of Motion," *Physical Review*, Vol. 33, No. 1, January 1961.

Wilczek, F., "Lectures on Black Hole Quantum Mechanics," *International Journal of Modern Physics A*, Vol. 13, No. 31, 1998.

Williams, G., "Scientists Charting the Undulating Rhythms of Superstrings Discuss Life in the Higher Dimensions," *Omni*, May 1987.

Winterberg, F., "Vortex Gerdien Arc with Magnetic Thermal Insulation," *Verlag der Zeitschrift fur Naturforschung*, Vol. 41a, March 1988.

Witten, E., "Reflections on the Fate of Spacetime," *Physics Today*, April 1996.

Wolff, M., "Fundamental Laws, Microphysics and Cosmology," *Physics Essays*, vol. 6, pp. 181-203, 1995.

Wolff, M., "Beyond the Point Particle - A Wave Structure for the Electron," Galilean Electrodynamics, Sept./Oct., 1995, pp. 83-91.

Wolff, M., "Exploring the Universe and the Origin of Its Laws," Temple University, *Frontier Perspectives*, Vol. 8, No.4,1997.

Wolff, M., "Origin of the Mysterious Instantaneous Transmission of Events in Science," *The Cosmic Light*, Vol. 4, No. 2, Spring 2002.

Y

Yam, P., "Current Events," *Scientific American*, December 1993.

Yurth, D. G., "A New Approach to Unified Field Theory," *Journal of New Energy*, Vol. 3, No. 2/3, 1998.

Yurth, D. G., Torsional Field Mechanics: Verification of Non-Local Field Effects in Human Biology," *Journal of New Energy*, Vol. 5, No. 2, 2000.

PART 2

Spiral Field Theory

CHAPTER 1

THE GEOMETRY OF THE TORIX

1.1 The definition of a torix

Torix (Fig. 1.1-1) is a particular case of the spiral field. The torix contains two spirals, one leading and one trailing. The leading spiral is represented by a part of a circle A_1 with radius r_1. One may think of a circle as a helical spiral whose wavelength λ_1 is equal to zero. The trailing spiral A_2 is one winding of a double toroidal spiral with the spiral radius r_2 and the wavelength λ_2. To construct the torix, we need one additional parameter called the *inversion radius of the spiral field r_i* (we will explain the physical meaning of this term in Chapter 2).

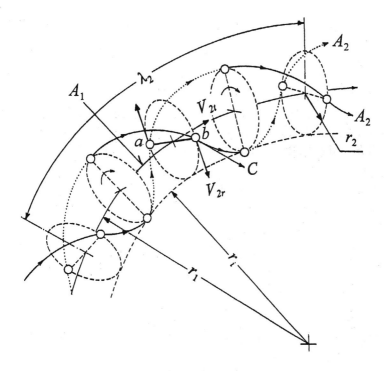

Fig. 1.1-1 The torix.

151

The geometry of the torix includes two features:

1. For a given set of torices with different radii r_1, the inversion radius of the spiral field r_i remains constant:

$$r_i = const. \tag{1.1-1}$$

2. The length of one winding of the torix trailing spiral L_2 is equal to the length of one winding of the leading spiral L_1:

$$L_2 = L_1 = 2\pi r_1 = 2\pi r_2 \left(\frac{r_1}{r_2} \right) \tag{1.1-2}$$

From Fig. 1.1-1, we obtain:

$$r_2 = r_1 - r_i \tag{1.1-3}$$

Since the inversion radius r_i is constant, it is convenient to express the parameters of the torix in relative terms as a function of the relative spiral radii b_1 and b_2 of the leading and trailing spirals respectively:

$$b_1 = \frac{r_1}{r_i} \tag{1.1-4}$$

$$b_2 = \frac{r_2}{r_i} \tag{1.1-5}$$

From Eqs. (1.1-3) - (1.1-5), we find:

$$b_2 = b_1 - 1 \tag{1.1-6}$$

The *torix vortex ratio* x_t is defined as the ratio of the spiral radius of the trailing spiral r_2 to the spiral radius of the leading spiral r_1. Considering Eq. (1.1-4) - (1.1-6), we obtain:

$$x_t = \frac{r_2}{r_1} = \frac{b_2}{b_1} = \frac{b_1 - 1}{b_1} \qquad (1.1\text{-}7)$$

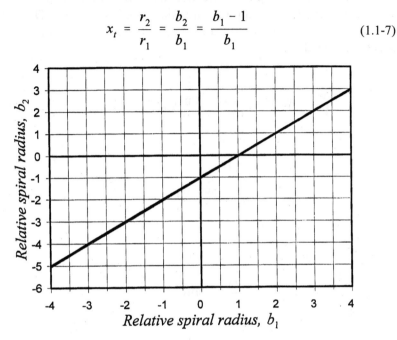

Fig. 1.1-2 Relative spiral radius b_2 as a function of the relative spiral radius b_1 - Eq. (1.1-6).

Fig. 1.1-2 shows a plot of Eq. (1.1-6). Relative spiral radius b_2 is linearly proportional to relative spiral radius b_1; it is equal to zero when $b_1 = 1$.

There are four main ranges of relative spiral radius b_1 within which the torices have their distinguishing characteristics:

1. Real negative range	$(1 \le b_1 < \infty)$
2. Real positive range	$(0.5 \le b_1 \le 1)$
3. Imaginary positive range	$(0 \le b_1 \le 0.5)$
4. Imaginary negative range	$(-\infty < b_1 \le 0)$.

Note that the adjacent ranges have common border values of relative spiral radius b_1.

1.2 Spirality

For both the leading and trailing spirals of the torix, we defined spirality by using three parameters, spiral slope, translational spirality, and rotational spirality. For the leading spiral, spiral slope angle $\varphi_1 = 90°$. Therefore, spiral slope γ_1, translational spirality γ_{1t}, and rotational spirality γ_{1r} are respectively equal to:

$$\gamma_1 = \frac{2\pi r_1}{\lambda_1} = \tan\varphi_1 \rightarrow \infty \qquad (1.2\text{-}1)$$

$$\gamma_{1t} = \frac{\lambda_1}{L_1} = \cos\varphi_1 = 0 \qquad (1.2\text{-}2)$$

$$\gamma_{1r} = \frac{2\pi r_1}{L_1} = \sin\varphi_1 = 1 \qquad (1.2\text{-}3)$$

For the trailing spiral, spiral slope γ_2, translational spirality γ_{2t}, and rotational spirality γ_{2r} are respectively equal to (Fig. 1.2-1):

$$\gamma_2 = \frac{2\pi r_2}{\lambda_2} = \tan\varphi_2 \qquad (1.2\text{-}4)$$

$$\gamma_{2t} = \frac{\lambda_2}{L_2} = \cos\varphi_2 \qquad (1.2\text{-}5)$$

$$\gamma_{2r} = \frac{2\pi r_2}{L_2} = \sin\varphi_2 \tag{1.2-6}$$

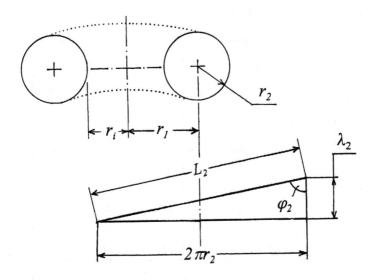

Fig. 1.2-1 Unwrapped winding of the trailing spiral of the torix.

From Eqs. (1.2-1), (1.2-3), (1.2-5), and (1.2-6) we obtain:

$$\gamma_{1t}^2 + \gamma_{1r}^2 = 1 \tag{1.2-7}$$

$$\gamma_{2t}^2 + \gamma_{2r}^2 = 1 \tag{1.2-8}$$

From Eq. (1.1-2), (1.1-4), (1.1-5), (1.1-7), and (1.2-6), we obtain the expression for rotational spirality γ_{2r} as a function of relative spiral radius b_1:

$$\gamma_{2r} = \frac{b_1 - 1}{b_1}$$
(1.2-9)

Eq. (1.2-9) is plotted in Fig. 1.2-2.

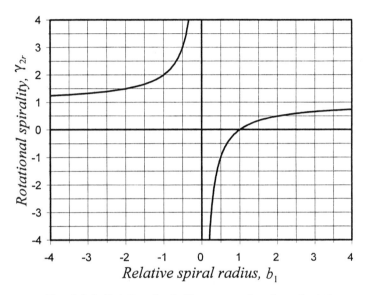

Fig. 1.2-2 Rotational spirality γ_{2r} as a function of relative
spiral radius b_1 - Eq. (1.2-9).

From Eqs. (1.2-8) and (1.2-9), we obtain the expression for
translational spirality γ_{2t} as a function of relative spiral radius b_1:

$$\gamma_{2t} = \frac{\sqrt{2b_1 - 1}}{b_1}$$
(1.2-10)

Eq. (1.2-10) is plotted in Fig. 1.2-3. Shown, with dotted lines in this graph,
are the plots for the parameters of the torix within imaginary ranges. Since
these parameters are expressed with imaginary numbers, their values must
be multiplied by i.

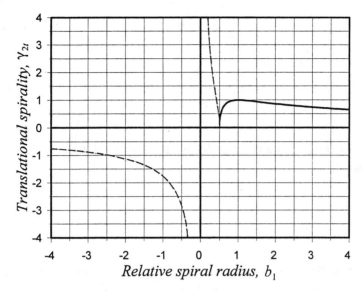

Fig. 1.2-3 Translational spirality γ_{2t} as a function of relative spiral radius b_1 - Eq. (1.2-10).

From Eqs. (1.2-9) and (1.2-10), we obtain the expression for spiral slope γ_2 as a function of relative spiral radius b_1:

$$\gamma_2 = \frac{\gamma_{2r}}{\gamma_{2t}} = \frac{b_1 - 1}{\sqrt{2b_1 - 1}} \qquad (1.2\text{-}11)$$

From Eqs. (1.1-2), (1.2-3), and (1.2-9), we find that the product of the rotational spiralities of the leading and trailing spirals of the torix is equal to:

$$\gamma_{1r}\gamma_{2r} = \frac{b_1 - 1}{b_1} \qquad (1.2\text{-}12)$$

1.3 Wavelength

From Eq. (1.2-2), we find that for the leading spiral, the wavelength λ_1 and the relative wavelength η_1 are equal to zero:

$$\lambda_1 = 0 \qquad\qquad (1.3\text{-}1)$$

$$\eta_1 = \frac{\lambda_1}{2\pi r_i} = 0 \qquad\qquad (1.3\text{-}2)$$

From Eq. (1.2-5), we find that for the trailing spiral, the wavelength λ_2 is equal to:

$$\lambda_2 = \gamma_{2t} L_2 \qquad\qquad (1.3\text{-}3)$$

Subsequently, from Eqs. (1.1-2), (1.1-4), (1.2-10), and (1.3-3), we obtain the equation for the wavelength λ_2 as a function of b_1:

$$\lambda_2 = 2\pi r_i \sqrt{2b_1 - 1} \qquad\qquad (1.3\text{-}4)$$

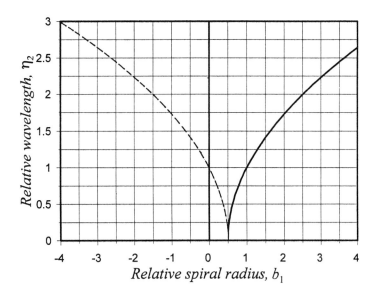

Fig. 1.3-1 Relative wavelength η_2 as a function of relative spiral radius b_1 - Eq. (1.3-5).

From Eq. (1.3-4), we find the equation for the relative wavelength η_2:

$$\eta_2 = \frac{\lambda_2}{2\pi r_i} = \sqrt{2b_1 - 1} \qquad (1.3\text{-}5)$$

Eq. (1.3-5) is plotted in Fig. 1.3-1.

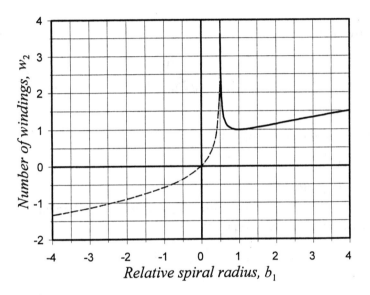

Fig. 1.4-1 The number of windings w_2 in the trailing spiral as a function of relative spiral radius b_1 - Eq. (1.4-3).

1.4 The number of windings

From Eq. (1.1-2) and Fig. 1.1-1, we find the number of windings of the leading spiral w_1 and trailing spiral w_2 of the torix:

$$w_1 = \frac{L_1}{\lambda_1} \qquad (1.4\text{-}1)$$

From Eqs. (1.1-2), (1.1-4), (1.2-10), (1.3-3), and (1.4-2) we derive the equation for the number of windings w_2 for the trailing spiral as a function

$$w_2 = \frac{L_1}{\lambda_2} = \frac{L_2}{\lambda_2} \tag{1.4-2}$$

of relative spiral radius b_1:

$$w_2 = \frac{1}{\gamma_{2t}} = \frac{b_1}{\sqrt{2b_1 - 1}} \tag{1.4-3}$$

Eq. (1.4-3) is plotted in Fig. 1.4-1.

1.5 Spiral density

For the leading spiral of the torix, spiral density ν_1 and relative spiral density ρ_1 are respectively equal to:

$$\nu_1 = \frac{1}{\lambda_1} \to \infty \tag{1.5-1}$$

$$\rho_1 = 2\pi r_i \nu_1 \to \infty \tag{1.5-2}$$

For the trailing spiral of the torix, spiral density ν_2 is equal to:

$$\nu_2 = \frac{1}{\lambda_2} \tag{1.5-3}$$

From Eqs. (1.3-4), and (1.5-3), we obtain the expression for spiral density v_2 as a function of relative spiral radius b_1:

$$v_2 = \frac{1}{2\pi r_i} \frac{1}{\sqrt{2b_1 - 1}} \tag{1.5-4}$$

From Eqs. (1.1-4), (1.4-3), and (1.5-4), we find another expression for spiral density v_2:

$$v_2 = \frac{w_2}{2\pi r_1} \tag{1.5-5}$$

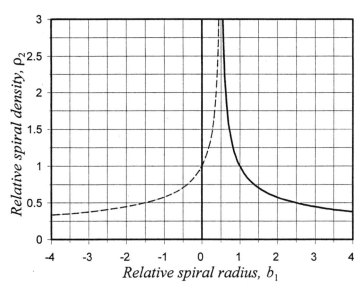

Fig. 1.5-1 Relative spiral density ρ_2 as a function of relative spiral radius b_1 - Eq. (1.5-6).

From Eq. (1.5-4), we find that for the trailing spiral relative spiral density ρ_2 is equal to:

$$\rho_2 = 2\pi r_i v_2 = \frac{1}{\sqrt{2b_1 - 1}} \tag{1.5-6}$$

Eq. (1.5-6) is plotted in Fig. 1.5-1.

CHAPTER 2

TYPES OF TORICES

2.1 Real and imaginary torices

The torices are divided into two types, real and imaginary. Both real and imaginary torices can be either negative or positive, as shown in Table 2.1-1.

Table 2.1-1 Ranges of relative spiral radius b_1 of real and imaginary torices.

Real torices		Imaginary torices	
Negative	Positive	Positive	Negative
$1 \leq b_1 < \infty$	$0.5 \leq b_1 \leq 1$	$0 \leq b_1 \leq 0.5$	$-\infty < b_1 \leq 0$

Real negative torices - These torices are located within the range $(1 \leq b_1 < +\infty)$ of the relative spiral radius b_1. Within this range, all the parameters of the torix are expressed with the real numbers. In the real negative torix (Fig. 2.1-1a), the trailing spiral is wound outside a circle with relative spiral radius $b_1 = 1$. This circle is called *Zero-Point-Energy (ZPE) ring*. The radius of the ZPE ring is equal to inversion radius r_i. The trailing spiral is assumed to be wound counter-clockwise as one looks along the direction of its translational propagation. A torix with infinite relative spiral radius b_1 has an infinite number of windings w_2, and has the appearance of a straight line. As the spiral radius of the torix decreases, the number of windings decreases (Fig. 2.1-1b). At $b_1 = 1$, the number of windings w_2 reduces to one, and the torix reduces to the ZPE ring (Fig. 2.1-1c).

Real positive torices - These torices are located within the range $(0.5 \leq b_1 \leq 1)$ of relative spiral radius b_1. Within this range, all torix parameters are expressed with real numbers. In the real positive torix (Fig. 2.1-2a), the trailing spiral is wound inside the ZPE field ring. The trailing spiral is assumed to be wound clockwise as one looks along the direction of its translational propagation. As relative spiral radius b_1 reduces, the number of windings w_2 increases (Fig. 2.1-2b). At $b_1 = 0.5$, the torix reduces to the *half ring* (Fig. 2.1-2c) that is located in the plane

163

perpendicular to the plane containing the ZPE ring. The half ring has an infinite number of windings w_2.

a) $b_1 = 2.3$

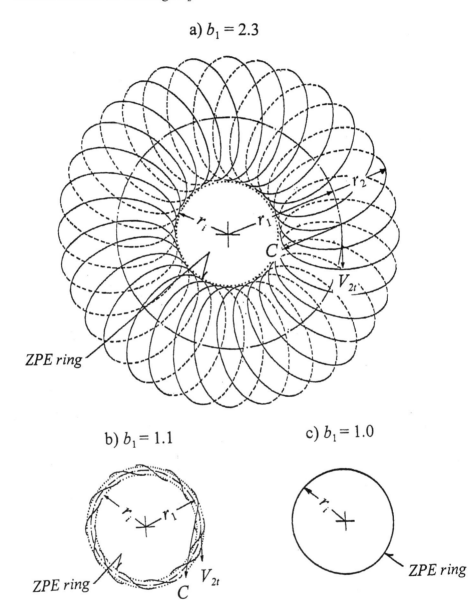

b) $b_1 = 1.1$ ### c) $b_1 = 1.0$

Fig. 2.1-1 Tori with real negative torices.

Shown in Figs. 2.1-1 through 2.1-4 are three components of the torix velocity:

V_{2t} = translational velocity of the torix trailing spiral

V_{2r} = rotational velocity of the torix trailing spiral

C = ultimate spiral field velocity.

The definition and physical meaning of these parameters will be discussed in Chapter 4.

a) $b_1 = 0.95$ b) $b_1 = 0.7$

c) $b_1 = 0.5$

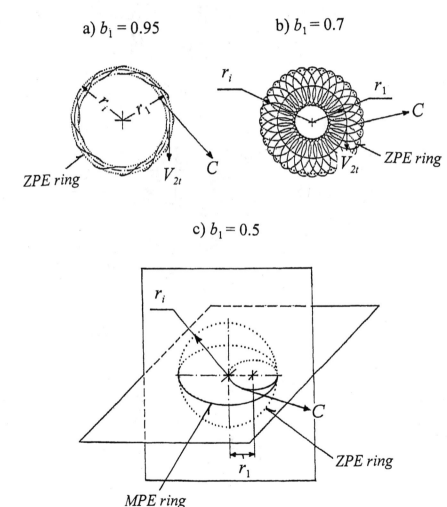

Fig. 2.1-2 Tori with real positive torices.

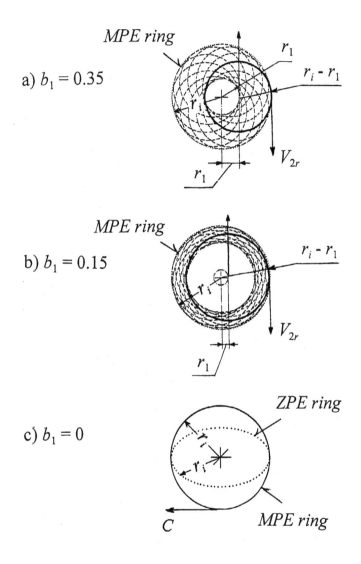

a) $b_1 = 0.35$

MPE ring

r_1

$r_i - r_1$

r_i

V_{2r}

r_1

b) $b_1 = 0.15$

MPE ring

$r_i - r_1$

r_i

V_{2r}

r_1

c) $b_1 = 0$

ZPE ring

r_i

r_i

C

MPE ring

Fig. 2.1-3 Tori with imaginary positive torices.

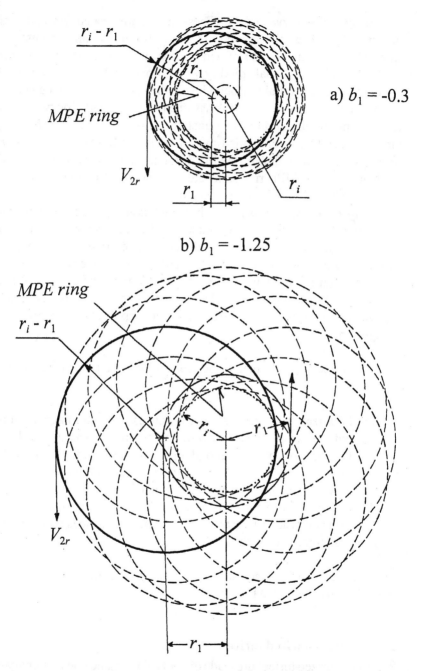

Fig. 2.1-4 Tori with imaginary negative torices.

Imaginary positive torices - These torices are located within the range $(0 \leq b_1 \leq 0.5)$ of relative spiral radius b_1. Within this range, translational spirality γ_{2t} and the number of windings w_2 are expressed with imaginary numbers, while rotational spirality γ_{2r} continues to have real values. The imaginary trailing spiral is wound inside the inversion field ring (Fig. 2.1-3a, b). This imaginary spiral is assumed to be wound clockwise as one looks along the direction of its translational propagation. At $b_1 = 0$, the torix reduces to a circle called the *Maximum-Point-Energy (MPE) ring* (Fig. 2.1-3c). The radius of the MPE ring is equal to the inversion radius r_i. Note that the MPE ring is located on the plane perpendicular to the plane on which the ZPE ring is located.

Imaginary negative torices - These torices are located within the range $(-\infty < b_1 \leq 0)$ of relative spiral radius b_1. Within this range, translational spirality γ_{2t} and the number of windings w_2 are expressed with imaginary numbers, while rotational spirality γ_{2r} continues to have real values. The imaginary trailing spiral is wound outside the inversion field ring (Fig. 2.1-4a, b) . It is wound counter-clockwise as one looks along the direction of its translational propagation. At $b_1 \rightarrow \infty$, the torix has the appearance of a straight line.

The torices are assumed to be created by three kinds of polarization processes:

1) Polarization of the first kind produces real and imaginary torices with mutually reverse flow of energy, inward and outward
2) Polarization of the second kind further divides both real and imaginary torices into pairs of torices with mutually inverse vorticity, left-handed and right-handed
3) Polarization of the third kind further splits the torices with the same direction of vorticity into the pairs of torices with spiralities that complement each other.

Polarized pairs of torices are called *matched torices*. Their special properties are described below.

2.2 Reality-polarized torices

Reality-polarized torices are made of one real torix and one imaginary torix with the same direction of vorticity, either positive or negative. For negative reality-polarized torices (Fig. 2.2-1), the sum of real and imaginary rotational spiralities γ_{2r} and $\bar{\gamma}_{2r}$ is equal to:

$$\gamma_{2r} + \bar{\gamma}_{2r} = 2 \qquad (2.2\text{-}1)$$

From Eqs. (1.2-9) and (2.2-1), we find that for negative real and imaginary torices, respective relative spiral radii b_1 and $\bar{b}_1$ are related by:

$$\bar{b}_1 = -b_1 \qquad (2.2\text{-}2)$$

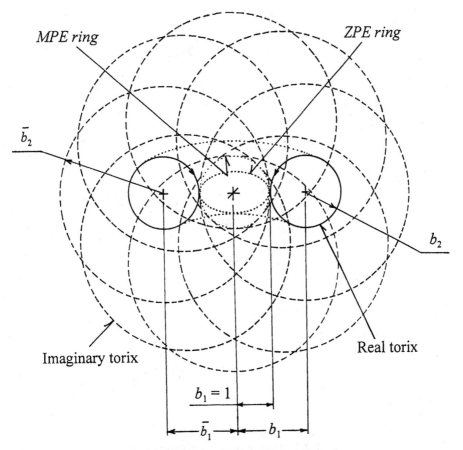

Fig. 2.2-1 Negative reality-polarized torices.

For positive reality-polarized torices (Fig. 2.2-2), the sum of real and imaginary rotational spiralities γ_{2r} and $\bar{\gamma}_{2r}$ is equal to:

$$\gamma_{2r} + \bar{\gamma}_{2r} = -2 \qquad (2.2\text{-}3)$$

From Eqs. (1.2-9) and (2.2-3), we find that for positive real and imaginary torices, respective relative spiral radii b_1 and $\bar{b}_1$ are related by:

$$\frac{1}{\bar{b}_1} + \frac{1}{b_1} = 4 \qquad (2.2\text{-}4)$$

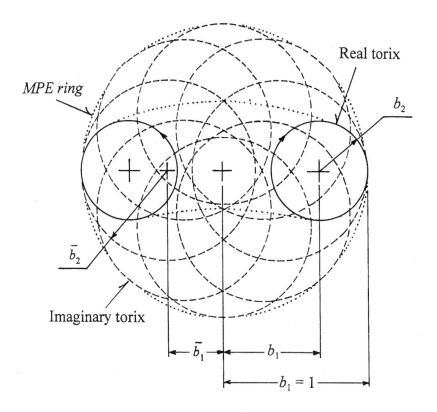

Fig. 2.2-2 Positive reality-polarized torices.

2.3 Vorticity-polarized torices

Vorticity-polarized torices have the same reality, either real (Fig. 2.3-1) or imaginary (Fig. 2.3-2), but opposite vorticity. Rotational spiralities γ_{2r}^- and γ_{2r}^+ of these torices have the same absolute values but opposite signs. Translational spiralities γ_{2t}^- and γ_{2t}^+ are equal to each other as given by:

$$\gamma_{2r}^- = -\gamma_{2r}^+ \qquad (2.3\text{-}1)$$

$$\gamma_{2t}^- = \gamma_{2t}^+ \qquad (2.3\text{-}2)$$

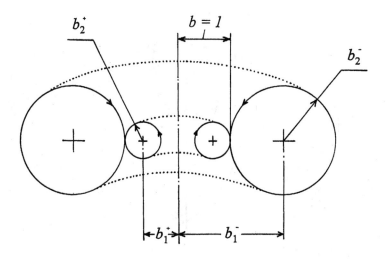

Fig. 2.3-1 Real vorticity-polarized torices.

It follows from Eqs. (1.2-9) and (2.3-1) that relative spiral radii b_1^- and b_1^+ of torices with inverse vorticity are related to each other by the equation:

$$\frac{1}{b_1^-} + \frac{1}{b_1^+} = 2 \qquad (2.3\text{-}3)$$

From Eq. (2.3-3), we obtain the relationship between relative spiral radii b_1^- and b_1^+ plotted in Fig. 2.3-3:

$$b_1^+ = -\frac{b_1^-}{2\,b_1^- - 1} \qquad\qquad (2.3\text{-}4)$$

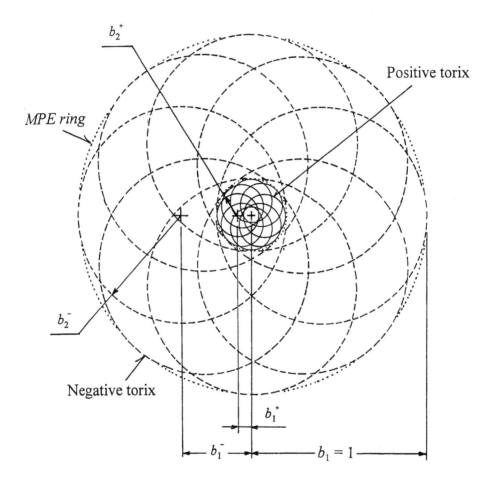

Fig. 2.3-2 Imaginary vorticity-polarized torices.

From Eq. (1.1-6), and (2.3-4), we find the relationship between spiral radii b_2^- and b_2^+ of vorticity-polarized torices:

$$b_2^+ = - \frac{b_2^-}{2b_2^- + 1}$$ (2.3-5)

From Eq. (2.3-3), we obtain:

$$b_1^- = \frac{b_1^+}{2b_1^+ - 1}$$ (2.3-6)

From Eqs. (1.1-6) and (2.3-6), we obtain:

$$b_2^- = - \frac{b_2^+}{2b_2^+ + 1}$$ (2.3-7)

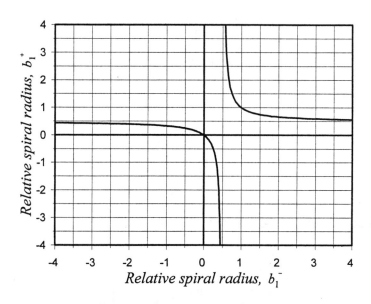

Fig. 2.3-3 The relationship between the relative spiral radii b_1^+ and b_1^- of vorticity-polarized torices - Eq. (2.3-4).

2.4 Complementary-polarized torices

Complementary-polarized torices have the same reality (either real or imaginary) and with the same vorticity (either positive or negative), but their rotational spiralities are complementary, as shown below.

Negative complementary-polarized torices (Fig. 2.4-1) - For negative complementary-polarized torices, the sum of rotational spiralities of inner and outer torices γ'_{2r} and γ''_{2r} is equal to:

$$\gamma'_{2r} + \gamma''_{2r} = 1 \qquad (2.4\text{-}1)$$

From Eqs. (1.2-9) and (2.4-1), we find the relationship between relative spiral radii b'_1 and b''_1 of negative complementary-polarized torices (Fig. 2.4-2):

$$\frac{1}{b'_1} + \frac{1}{b''_1} = 1 \qquad (2.4\text{-}2)$$

$$b'_1 = \frac{b''_1}{b''_1 - 1} \qquad (2.4\text{-}3)$$

$$b''_1 = \frac{b'_1}{b'_1 - 1} \qquad (2.4\text{-}4)$$

From Eqs. (1.1-6), (2.4-3), and (2.4-4), we obtain:

$$b'_1 - 1 = \frac{1}{b''_1 - 1} \qquad (2.4\text{-}5)$$

$$b_2' = \frac{1}{b_2''}$$

(2.4-6)

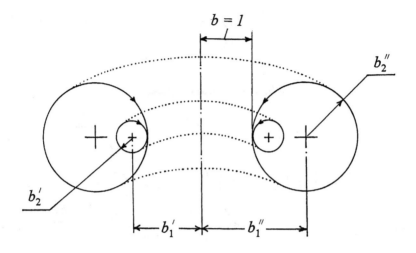

Fig. 2.4-1 Real negative complementary-polarized torices.

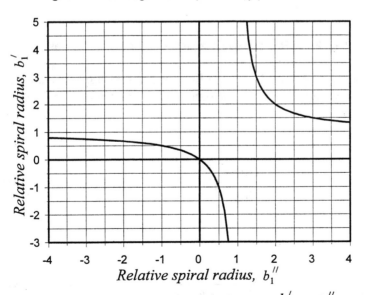

Fig. 2.4-2 The relationship between relative spiral radii b_1' and b_1'' of inner and outer negative complementary-polarized torices - Eq. (2.4-3).

Positive complementary-polarized torices - From Eqs. (2.3-1) and (2.4-1), we find that for positive complementary-polarized torices, the sum of rotational spiralities γ'_{2r} and γ''_{2r} of inner and outer torices is equal to (Fig. 2.4-3):

$$\gamma'_{2r} + \gamma''_{2r} = -1 \qquad (2.4\text{-}7)$$

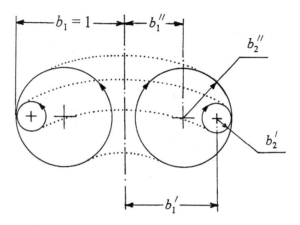

Fig. 2.4-3 Real positive complementary-polarized torices.

From Eqs. (1.2-9) and (2.4-7), we find the relationship between relative spiral radii b'_1 and b''_1 of inner and outer positive complementary-polarized torices (Fig. 2.4-4):

$$\frac{1}{b'_1} + \frac{1}{b''_1} = 3 \qquad (2.4\text{-}8)$$

$$b'_1 = \frac{b''_1}{3b''_1 - 1} \qquad (2.4\text{-}9)$$

$$b_1'' = \frac{b_1'}{3b_1' - 1} \qquad (2.4\text{-}10)$$

From Eqs. (1.1-6) and (2.4-9), we obtain:

$$b_2' = - \frac{2b_1'' - 1}{3b_1'' - 1} \qquad (2.4\text{-}11)$$

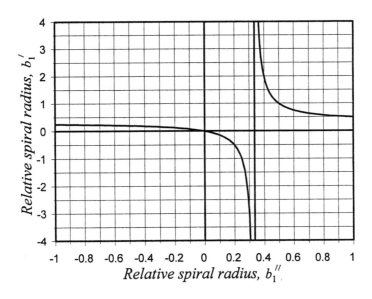

Fig. 2.4-4 The relationship between relative spiral radii b_1' and b_1'' of inner and outer positive complementary-polarized torices - Eq. (2.4-9).

CHAPTER 3

THE GEOMETRY OF THE HELIX

3.1 The definition of the helix

The helix (Fig. 3.1-1) is a particular case of the spiral field. It contains two spirals, leading and trailing. The leading spiral A_3 is one winding of a double helical spiral with the spiral radius r_3 and the wavelength λ_3. The trailing spiral is one winding of a double helical spiral A_4 with the spiral radius r_4 and the wavelength λ_4. It is wound around the leading helical spiral A_3.

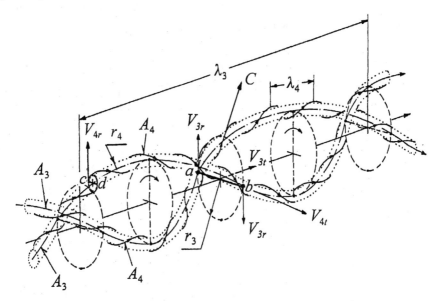

Fig. 3.1-1 The helix.

Shown on Figs. 3.1-1 are the three velocity components of both the leading and trailing spirals:

V_{3t} = translational velocity of the helix leading spiral

V_{3r} = rotational velocity the helix leading spiral
V_{4t} = translational velocity of the helix trailing spiral
V_{4r} = rotational velocity the helix trailing spiral
C = ultimate spiral field velocity.

Helix geometry includes two features:

1. For a given set of helices with different spiral radii r_3, the inversion radius r_i remains constant:

$$r_i = const. \tag{3.1-1}$$

2. The length of one winding of trailing helical spiral L_4 is equal to the length of leading helical spiral L_3 (Fig. 3.1-2):

$$L_4 = L_3 = 2\pi r_3 \left(\frac{r_3}{r_4}\right) = 2\pi r_4 \left(\frac{r_3}{r_4}\right)^2 \tag{3.1-2}$$

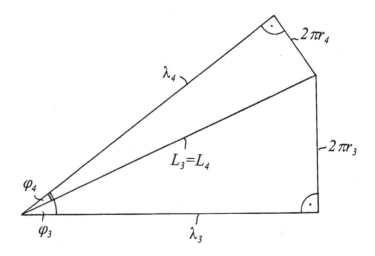

Fig. 3.1-2 Unwrapped windings of the helix.

The radius of helix trailing spiral r_4 is equal to:

$$r_4 = r_3 - r_i \tag{3.1-3}$$

Fig. 3.1-3 shows a plot of Eq. (3.1-6).

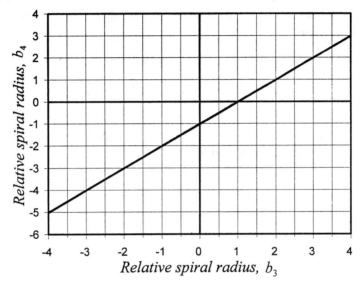

Fig. 3.1-3 Relative spiral radius b_4 as a function of relative spiral radius b_3 - Eq. (3.1-6).

Since the inversion radius r_i is constant, it is convenient to express the parameters in relative terms as a function of relative spiral radii b_3 and b_4:

$$b_3 = \frac{r_3}{r_i} \tag{3.1-4}$$

$$b_4 = \frac{r_4}{r_i} \tag{3.1-5}$$

From Eqs. (3.1-3) - (3.1-5), we obtain:

$$b_4 = b_3 - 1 \qquad (3.1\text{-}6)$$

The *helix vortex ratio* x_h is defined as the ratio of the spiral radius of the trailing spiral r_4 to the spiral radius of the leading spiral r_3. Considering Eq. (3.1-4) - (3.1-6), we obtain:

$$x_h = \frac{r_4}{r_3} = \frac{b_4}{b_3} = \frac{b_3 - 1}{b_3} \qquad (3.1\text{-}7)$$

By considering Eqs. (3.1-2) - (3.1-4), we can express spiral length L_3 as a function of relative spiral radius b_3:

$$L_3 = L_4 = 2\pi r_i \frac{b_3^2}{b_3 - 1} \qquad (3.1\text{-}8)$$

3.2 Wavelength

From Fig. (3.1-2), we find that for the leading spiral, the wavelength λ_3 is equal to:

$$\lambda_3 = \sqrt{L_3^2 - (2\pi r_3)^2} \qquad (3.2\text{-}1)$$

Subsequently, from Eq. (3.1-4), (3.1-8), and (3.2-1), we obtain the equation for the wavelength λ_3 and relative wavelength η_3 as a function of b_3:

$$\lambda_3 = 2\pi r_i \frac{b_3}{b_3 - 1} \sqrt{2b_3 - 1} \qquad (3.2\text{-}2)$$

$$\eta_3 = \frac{\lambda_3}{2\pi r_i} = \frac{b_3}{b_3 - 1}\sqrt{2b_3 - 1} \qquad (3.2\text{-}3)$$

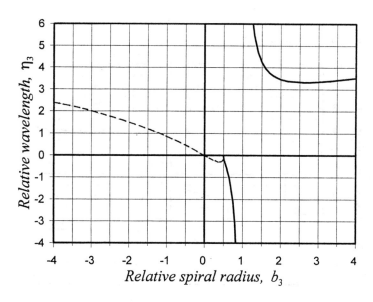

Fig. 3.2-1 Relative wavelength η_3 as a function of relative spiral radius b_3 - Eq. (3.2-3).

Eq. (3.2-3) is plotted in Fig. 3.2-1.

From Fig. (3.1-2), we find that for the trailing spiral, the wavelength λ_4 is equal to:

$$\lambda_4 = \sqrt{L_4^2 - (2\pi r_4)^2} \qquad (3.2\text{-}4)$$

Subsequently, from Eq. (3.1-5), (3.1-6), and (3.1-8), we obtain the equation for the wavelength λ_4 and relative wavelength η_4 as a function of b_3:

$$\lambda_4 = \frac{2\pi r_i}{b_3 - 1}\sqrt{b_3^4 - (b_3 - 1)^4} \qquad (3.2\text{-}5)$$

$$\eta_4 = \frac{\lambda_4}{2\pi r_i} = \frac{\sqrt{b_3^{\,4} - (b_3 - 1)^4}}{b_3 - 1} \qquad (3.2\text{-}6)$$

Eq. (3.2-6) is plotted in Fig. 3.2-2.

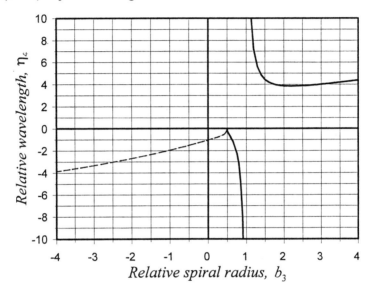

Fig. 3.2-2 Relative wavelength η_4 as a function of the relative spiral radius b_3 - Eq. (3.2-6).

3.3 Spirality

For both the leading and trailing spirals of the helix, we defined spirality by using three parameters, spiral slope γ_3, translational spirality γ_{3t}, and rotational spirality γ_{3r} (Fig. 3.1-2). For the leading spiral, these parameters are respectively equal to:

$$\gamma_3 = \frac{2\pi r_3}{\lambda_3} = \tan\varphi_3 \qquad (3.3\text{-}1)$$

$$\gamma_{3t} = \frac{\lambda_3}{L_3} = \cos\varphi_3 \qquad (3.3\text{-}2)$$

$$\gamma_{3r} = \frac{2\pi r_3}{L_3} = \sin\varphi_3 \qquad (3.3\text{-}3)$$

For the trailing spiral, spiral slope γ_4, translational spirality γ_{4t}, and rotational spirality γ_{4r} are respectively equal to:

$$\gamma_4 = \frac{2\pi r_4}{\lambda_4} = \tan\varphi_4 \qquad (3.3\text{-}4)$$

$$\gamma_{4t} = \frac{\lambda_4}{L_4} = \cos\varphi_4 \qquad (3.3\text{-}5)$$

$$\gamma_{4r} = \frac{2\pi r_4}{L_4} = \sin\varphi_4 \qquad (3.3\text{-}6)$$

From Eqs. (3.3-2), (3.3-3), (3.3-5), and (3.3-6), we obtain:

$$\gamma_{3t}^2 + \gamma_{3r}^2 = 1 \qquad (3.3\text{-}7)$$

$$\gamma_{4t}^2 + \gamma_{4r}^2 = 1 \qquad (3.3\text{-}8)$$

From Eqs. (3.1-8), (3.2-2), and (3.3-2), we derive the equation for translational spirality γ_{3t} as a function of b_3:

$$\gamma_{3t} = \frac{\sqrt{2b_3 - 1}}{b_3} \qquad (3.3\text{-}9)$$

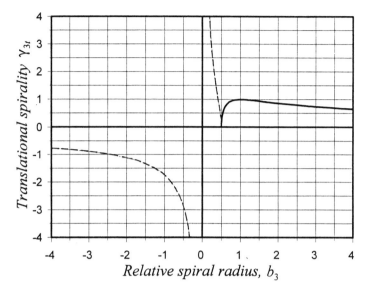

Fig. 3.3-1 Translational spirality γ_{3t} as a function of relative spiral radius b_3 - Eq. (3.3-9).

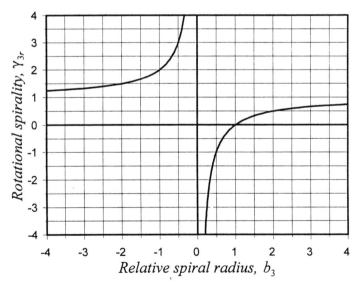

Fig. 3.3-2 Rotational spirality γ_{3r} as a function of relative spiral radius b_3 - Eq. (3.3-10).

From Eq. (3.1-4), (3.1-8), and (3.3-3), we derive the equation for rotational spirality γ_{3r} as a function of b_3:

$$\gamma_{3r} = \frac{b_3 - 1}{b_3} \qquad (3.3\text{-}10)$$

Eqs. (3.3-9) and (3.3-10) are plotted in Figs. 3.3-1 and 3.3-2.

From Eqs. (3.1-8), (3.2-5), and (3.3-5), we derive the equation for translational spirality γ_{4t} as a function of b_3:

$$\gamma_{4t} = \frac{1}{b_3^2} \sqrt{b_3^4 - (b_3 - 1)^4} \qquad (3.3\text{-}11)$$

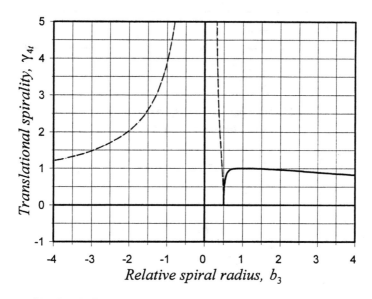

Fig. 3.3-3 Translational spirality γ_{4t} as a function of relative spiral radius b_3 - Eq. (3.3-11).

Eq. (3.3-11) is plotted in Fig. 3.3-3.

From Eqs. (3.1-5), (3.1-6), (3.1-8), and (3.3-6), we derive the equation for the rotational spirality γ_{4r} as a function of b_3:

$$\gamma_{4r} = \left(\frac{b_3 - 1}{b_3} \right)^2 \qquad (3.3\text{-}12)$$

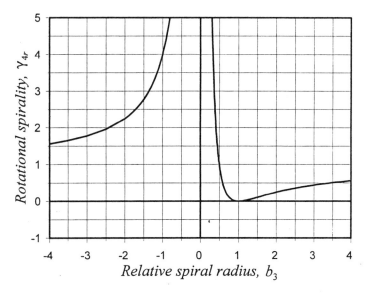

Fig. 3.3-4 Rotational spirality γ_{4r} as a function of relative spiral radius b_3 - Eq. (3.3-12).

Eq. (3.3-12) is plotted in Fig. 3.3-4.

From Eqs. (3.3-10) and (3.3-12), we obtain:

$$\gamma_{3r}\gamma_{4r} = \left(\frac{b_3 - 1}{b_3} \right)^3 \qquad (3.3\text{-}13)$$

3.4 The number of windings

The number of windings w_3 along the length L_3 of the leading spiral of the helix is equal to:

$$w_3 = \frac{L_3}{\lambda_3} \qquad (3.4\text{-}1)$$

From Eqs. (3.3-2), (3.3-9), and (3.4-1), we derive the equation for the number of windings w_3 as a function of relative spiral radius b_3:

$$w_3 = \frac{1}{\gamma_{3t}} = \frac{b_3}{\sqrt{2b_3 - 1}} \qquad (3.4\text{-}2)$$

Eq. (3.4-2) is plotted in Fig. 3.4-1.

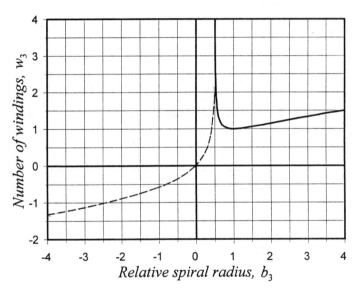

Fig. 3.4-1 The number of windings w_3 in the helix leading spiral as a function of relative spiral radius b_3 - Eq. (3.4-2).

Similarly to Eq. (1.4-2), a number of windings w_4 in the helix trailing spiral is equal to:

$$w_4 = \frac{L_4}{\lambda_4} \qquad (3.4\text{-}3)$$

From Eqs. (3.3-5), (3.3-11), and (3.4-3), we derive the equation for the number of windings w_4 as a function of b_3:

$$w_4 = \frac{1}{\gamma_{4t}} = \frac{b_3^2}{\sqrt{b_3^4 - (b_3 - 1)^4}} \qquad (3.4-4)$$

Eq. (3.4-4) is plotted in Fig. 3.4-2.

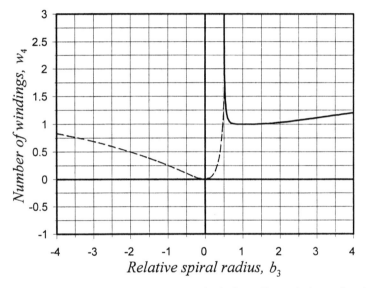

Fig. 3.4-2 The number of windings w_4 in the helix trailing spiral as a function of relative spiral radius b_3 - Eq. (3.4-4).

3.5 Spiral density

For the helix leading spiral, spiral density v_3 is equal to:

$$v_3 = \frac{1}{\lambda_3} \qquad (3.5-1)$$

From Eqs. (3.2-2) and (3.5-1), we derive the equations for spiral density v_3 and relative spiral density ρ_3 as a function of relative spiral radius b_3:

$$v_3 = \frac{1}{2\pi r_i} \frac{b_3 - 1}{b_3\sqrt{2b_3 - 1}} \qquad (3.5\text{-}2)$$

$$\rho_3 = v_3 2\pi r_i = \frac{b_3 - 1}{b_3\sqrt{2b_3 - 1}} \qquad (3.5\text{-}3)$$

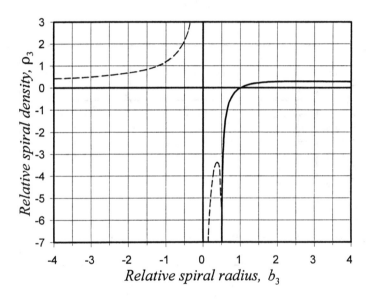

Fig. 3.5-1 Relative spiral density ρ_3 as a function of relative spiral radius b_3 - Eq. (3.5-3).

Eq. (3.5-3) is plotted in Fig. 3.5-1.
For the trailing spiral, spiral density v_4 is equal to:

$$v_4 = \frac{1}{\lambda_4} \qquad (3.5\text{-}4)$$

The Geometry of the Helix

From Eqs. (3.2-5) and (3.5-4), we derive the equations for spiral density v_4 and relative spiral density ρ_4 as a function of b_3:

$$v_4 = \frac{1}{2\pi r_i} \frac{b_3 - 1}{\sqrt{b_3^4 - (b_3 - 1)^4}} \qquad (3.5\text{-}5)$$

$$\rho_4 = v_4 2\pi r_i = \frac{b_3 - 1}{\sqrt{b_3^4 - (b_3 - 1)^4}} \qquad (3.5\text{-}6)$$

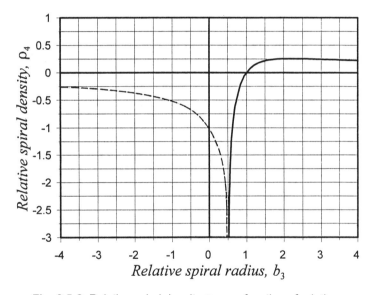

Fig. 3.5-2 Relative spiral density ρ_4 as a function of relative spiral radius b_3 - Eq. (3.5-6).

Eq. (3.5-6) is plotted in Fig. 3.5-2.

CHAPTER 4

PHYSICAL PROPERTIES OF TORICES

4.1 Velocities

In the torix leading spiral, spiral velocity V_1 is equal to the geometrical sum of translational velocity V_{1t} and rotational velocity V_{1r}:

$$V_{1t}^2 + V_{1r}^2 = V_1^2 \qquad (4.1-1)$$

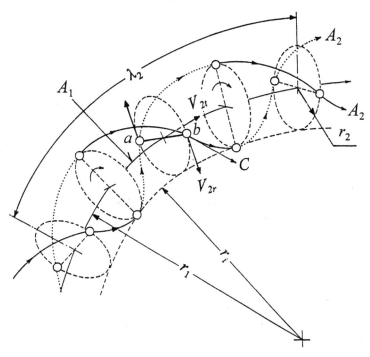

Fig. 4.1-1 Torix velocities.

Similarly, in the torix trailing spiral, spiral velocity V_2 is equal to the geometrical sum of translational velocity V_{2t} and rotational velocity V_{2r}: (Fig. 4.1-1):

193

$$V_{2t}^2 + V_{2r}^2 = V_2^2 \qquad (4.1\text{-}2)$$

The next three equations describe specific relationships between the velocities of the leading and trailing spirals that form the torix. Leading spiral velocity V_1 is equal to translational velocity V_{2t} of the trailing spiral:

$$V_1 = V_{2t} \qquad (4.1\text{-}3)$$

Trailing spiral velocity V_2 is greater than leading spiral velocity V_1 and is assumed to be equal to the ultimate spiral field velocity C:

$$V_2 = C \qquad (4.1\text{-}4)$$

Since the leading spiral is a circle, its translational velocity V_{1t} is equal to zero:

$$V_{1t} = 0 \qquad (4.1\text{-}5)$$

From Eqs. (4.1-1), (4.1-3), and (4.1-5) we obtain:

$$V_{1r} = V_{2t} \qquad (4.1\text{-}6)$$

From Eqs. (4.1-2), and (4.1-4), we find that for the trailing spiral the geometrical sum of translational velocity V_{2t} and rotational velocity V_{2r} is equal to the ultimate spiral field velocity C:

$$V_{2t}^2 + V_{2r}^2 = C^2 \qquad (4.1\text{-}7)$$

For the sake of convenience, we will express torix relative velocities β_{1t}, β_{1r}, β_{2t}, and β_{2r}, by the following equations:

$$\beta_{1t} = \frac{V_{1t}}{C} = 0 \qquad (4.1\text{-}8)$$

$$\beta_{1r} = \frac{V_{1r}}{C} \qquad (4.1\text{-}9)$$

$$\beta_{2t} = \frac{V_{2t}}{C} \qquad (4.1\text{-}10)$$

$$\beta_{2r} = \frac{V_{2r}}{C} \qquad (4.1\text{-}11)$$

The equations below provide the relationships between torix relative velocities. From Eqs. (4.1-6) and (4.1-8) - (4.1-10), we obtain:

$$\beta_{1t}^2 + \beta_{1r}^2 = \beta_{2t}^2 \qquad (4.1\text{-}12)$$

From Eqs. (4.1-7), (4.1-10), and (4.1-11), we obtain:

$$\beta_{2t}^2 + \beta_{2r}^2 = 1 \qquad (4.1\text{-}13)$$

The equations below express torix relative velocities as a function of relative torix radius b_1. From Eqs. (1.2-5), (1.2-10), (4.1-6), (4.1-9), and (4.1-10), we obtain:

$$\beta_{1r} = \beta_{2t} = \gamma_{2t} = \frac{\sqrt{2b_1 - 1}}{b_1} \qquad (4.1\text{-}14)$$

From Eqs. (1.1-7), (1.2-9), and (4.1-11), we obtain:

$$\beta_{2r} = \gamma_{2r} - \frac{b_1 - 1}{b_1} - x_t \qquad (4.1\text{-}15)$$

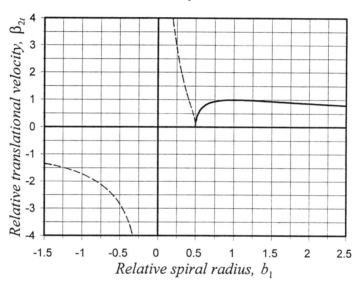

Fig. 4.1-2 Relative translational velocity β_{2t} as a function of relative spiral radius b_1 - Eq. (4.1-14).

Figs. 4.1-2 and 4.1-3 show plots of Eqs. (4.1-14) and (4.1-15). Note that the maximum value of relative translational velocity $\beta_{2t} = 1$ is reached at $b_1 = 1$.

From Eqs. (4.1-13) and (4.1-15), the relative radius of the leading spiral b_1 is equal to:

$$b_1 = \frac{1}{1 - \beta_{2r}} = \frac{1 + \beta_{2r}}{\beta_{2t}^2} \qquad (4.1\text{-}16)$$

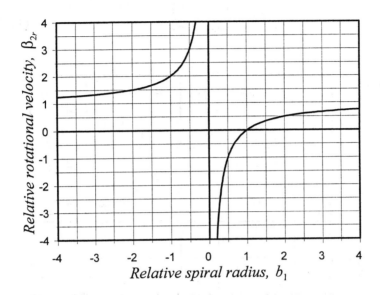

Fig. 4.1-3 Relative rotational velocity β_{2r} as a function of relative spiral radius b_1
- Eq. (4.1-15).

4.2 Spiral frequencies and densities

By definition, the fundamental spiral field frequency f_o is equal to:

$$f_o = \frac{C}{2\pi r_i} \qquad (4.2\text{-}1)$$

Considering Eqs. (1.1-3), (1.1-4), and (4.1-11), the frequencies of the torix leading and trailing spirals f_1 and f_2 are respectively given by:

$$f_1 = \frac{V_{1r}}{2\pi r_1} \qquad (4.2\text{-}2)$$

$$f_2 = \frac{V_{2r}}{2\pi r_2} = \frac{C}{2\pi r_1} \qquad (4.2\text{-}3)$$

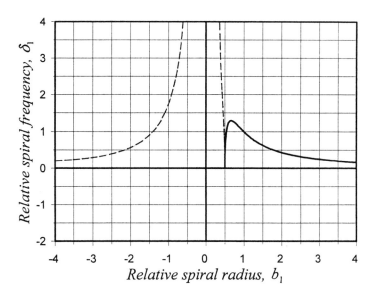

Fig. 4.2-1 Relative spiral frequency δ_1 as a function of relative spiral radius b_1 - Eq. (4.2-4).

From Eqs. (1.1-4), (4.1-9), (4.1-14), (4.2-1), and (4.2-2), we derive the equation for the relative spiral frequency of the torix leading spiral δ_1:

$$\delta_1 = \frac{f_1}{f_o} = \frac{\sqrt{2b_1 - 1}}{b_1^2} \qquad (4.2\text{-}4)$$

From Eqs. (1.1-5), (1.1-6), (4.1-11), (4.1-15), (4.2-1), and (4.2-3), we derive the equation for the relative spiral frequency of the torix trailing spiral δ_2:

$$\delta_2 = \frac{f_2}{f_o} = \frac{1}{b_1} \qquad (4.2\text{-}5)$$

Figs. 4.2-1 and 4.2-2 show plots of Eqs. (4.2-4) and (4.2-5). Note that in Fig. 4.2-1 the maximum value of relative spiral frequency δ_1 is reached at $b_1 = 2/3$.

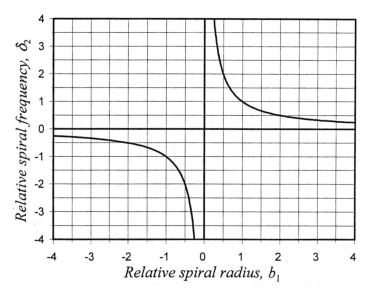

Fig. 4.2-2 Torix relative spiral frequency δ_2 as a function of relative spiral radius b_1 - Eq. (4.2-5).

By definition and from Eq. (4.1-5), spiral densities of the torix leading and trailing spirals v_1 and v_2 are respectively equal to:

$$v_1 = \frac{f_1}{V_{1t}} \rightarrow \infty \qquad (4.2\text{-}6)$$

$$v_2 = \frac{f_2}{V_{2t}} \qquad (4.2\text{-}7)$$

With Eqs. (4.1-10), (4.1-14), (4.2-5), and (4.2-7), we can express the torix spiral density v_2 (spiral density of the torix trailing spiral) as a function of the relative spiral radius b_1:

$$v_2 = \frac{f_o}{C} \frac{1}{\sqrt{2b_1 - 1}} \qquad (4.2\text{-}8)$$

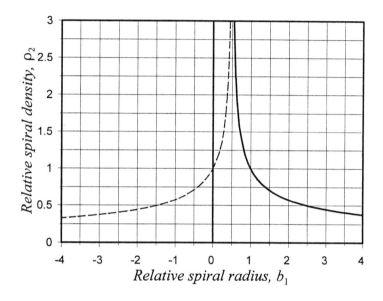

Fig. 4.2-3 Torix relative spiral density ρ_2 as a function of relative spiral radius b_1 - Eq. (4.2-9).

From Eq. (4.2-8), we obtain the equation for torix relative spiral density ρ_2:

$$\rho_2 = v_2 \frac{C}{f_o} = \frac{1}{\sqrt{2b_1 - 1}} \qquad (4.2\text{-}9)$$

Fig. 4.2-3 shows a plot of Eq. (4.2-9).

4.3 Gravitational mass

Torix relative gravitational mass m_{tg}/m_o is a ratio of torix gravitational mass m_{tg} to the rest mass of reality-polarized torices m_o, and is equal to:

$$\frac{m_{tg}}{m_o} = abs\left(\frac{r_2}{2r_1}\right) = abs\left(\frac{x_t}{2}\right) \qquad (4.3\text{-}1)$$

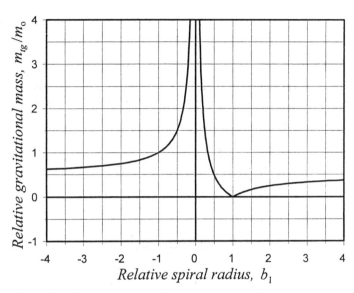

Fig. 4.3-1 Torix relative gravitational mass m_{tg}/m_o as a function of relative spiral radius b_1 - Eq. (4.3-3).

From Eqs.(4.1-13), (4.1-15), and (4.3-1), we can express torix relative gravitational mass m_{tg}/m_o as functions of relative velocities β_{2r}, β_{2t} and relative spiral radius b_1:

$$\frac{m_{tg}}{m_o} = abs\left(\frac{\beta_{2r}}{2}\right) = \frac{\sqrt{1 - \beta_{2t}^2}}{2} \qquad (4.3\text{-}2)$$

$$\frac{m_{tg}}{m_o} = abs\left(\frac{b_1 - 1}{2b_1}\right) \qquad (4.3\text{-}3)$$

Fig. (4.3-1) shows a plot of Eq. (4.3-3).

4.4 Electric charge

Torix relative electric charge e_t/e_o is a ratio of torix electric charge e_t to the rest electric charge of reality-polarized torices e_o, and is equal to:

$$\frac{e_t}{e_o} = \pm abs \left(\frac{r_2}{2r_1} \right) = \pm abs \left(\frac{x_t}{2} \right) \qquad (4.4\text{-}1)$$

From Eqs.(4.1-13), (4.1-15), and (4.4-1), we can express torix relative electric charge e_t/e_o as a function of either relative velocities β_{2r}, β_{2t} and relative spiral radius b_1:

$$\frac{e_t}{e_o} = \pm\, abs \left(\frac{\beta_{2r}}{2} \right) = \pm \frac{\sqrt{1 - \beta_{2t}^2}}{2} \qquad (4.4\text{-}2)$$

$$\frac{e_t}{e_o} = \pm abs \left(\frac{b_1 - 1}{2b_1} \right) \qquad (4.4\text{-}3)$$

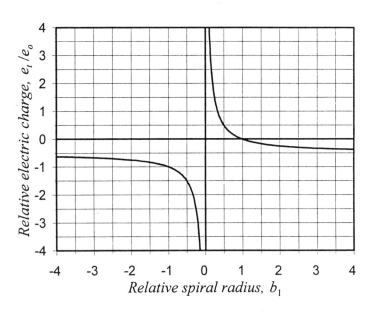

Fig. 4.4-1 Torix relative electric charge e_t/e_o as a function of relative spiral radius b_1 - Eq. (4.4-3).

Fig. (4.4-1) shows a plot of Eq. (4.4-3).

4.5 Inversion radius

We will derive below the equation for the inversion radius r_i by replacing the toroidal spiral field with a particle having electric charge e and inertial mass m_i. The particle is moving with the translational velocity V_{2t} around a stationary particle having the electric charge e_o and located at the distance r_1.

From Coulomb's law, electric force F_e acting on the moving particle with electric charge e is equal to:

$$F_e = - \frac{ke_o e}{r_1^2} \qquad (4.5\text{-}1)$$

From Newton's law and Eq. (4.1-10), inertial force F_i acting on the particle with inertial mass m_i moving with the translational velocity V_{2t} is equal to:

$$F_i = \frac{m_i V_{2t}^2}{r_1} = \frac{m_i C^2}{r_1} \beta_{2t}^2 \qquad (4.5\text{-}2)$$

From Eqs. (4.5-1) and (4.5-2), at equilibrium condition $F_e = F_i$, we obtain:

$$r_1 = - \frac{ke_o e}{m_i C^2} \frac{1}{\beta_{2t}^2} \qquad (4.5\text{-}3)$$

When $\beta_{2t} \ll 1$, $m_i \approx m_o$ and $e \approx -e_o$ Therefore, Eq. (4.5-3) reduces to the form:

$$r_1 = \frac{ke_o^2}{2m_o C^2} \frac{2}{\beta_{2t}^2} \qquad (4.5\text{-}4)$$

When $\beta_{2t} \ll 1$, $\beta_{2r} \approx 1$, Therefore, from Eqs. (1.1-4) and (4.1-16), the spiral radius r_1 is given by:

$$r_1 = r_i \frac{2}{\beta_{2t}^2} \qquad (4.5\text{-}5)$$

Thus, from Eqs. (4.5-4) and (4.5-5), we obtain the equation for inversion spiral radius r_i:

$$r_i = \frac{ke_o^2}{2m_o C^2} \qquad (4.5\text{-}6)$$

From Eqs. (4.2-1) and (4.5-6), we find a new expression for fundamental spiral field frequency f_o:

$$f_o = \frac{m_o C^3}{k \pi e_o^2} \qquad (4.5\text{-}7)$$

4.6 Inertial mass

By definition, torix inertial mass m_{ti} and electric charge e_t are equal to:

$$m_{ti} = \frac{m_t}{2} \qquad (4.6\text{-}1)$$

$$e_t = \frac{e}{2} \qquad (4.6\text{-}2)$$

From Eqs. (4.5-3), (4.6-1), and (4.6-2), we find torix relative inertial mass as a function of spiral radius r_1 and relative translational velocity β_{2t}:

$$\frac{m_{ti}}{m_o} = -\frac{ke_o e_t}{m_o C^2} \frac{1}{r_1 \beta_{2t}^2} \tag{4.6-3}$$

By using Eqs. (1.1-4), (4.4-2), and (4.5-6), we can reduce Eq. (4.6-3) to the form:

$$\frac{m_{ti}}{m_o} = \frac{1}{b_1} \frac{\beta_{2r}}{\beta_{2t}^2} \tag{4.6-4}$$

Considering Eqs. (4.1.15) and (4.1-16), Eq. (4.6-4) takes the form:

$$\frac{m_{ti}}{m_o} = \frac{\beta_{2r}}{1 + \beta_{2r}} = \frac{x_t}{1 + x_t} \tag{4.6-5}$$

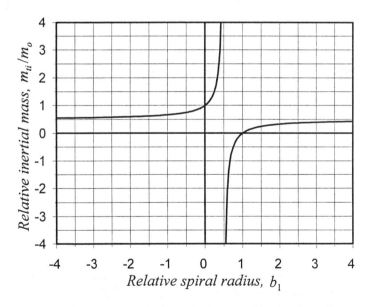

Fig. 4.6-1 Torix relative inertial mass m_{ti}/m_o as a function of relative spiral radius b_1 - Eq. (4.6-6).

From Eqs. (4.1-15) and (4.6-5), we find torix relative inertial mass to be a function of relative spiral radius b_1:

$$\frac{m_{ti}}{m_o} = \frac{b_1 - 1}{2b_1 - 1} \tag{4.6-6}$$

Fig. (4.6-1) shows a plot of Eq. (4.6-6).

4.7 Translational angular momentum

By definition, torix angular momentum P_{ta} is equal to:

$$P_{ta} = 2\pi r_1 m_{ti} V_{2t} \tag{4.7-1}$$

From Eqs. (1.1-4), (4.1-10), (4.1-14), (4.6-6), and (4.7-1), we obtain:

$$P_{ta} = 2\pi r_i m_o C \frac{b_1 - 1}{\sqrt{2b_1 - 1}} \tag{4.7-2}$$

From Eqs. (4.1-14), (4.1-15), and (4.7-2), we obtain:

$$P_{ta} = 2\pi r_i m_o C \frac{\beta_{2r}}{\beta_{2t}} \tag{4.7-3}$$

From Eqs. (4.5-6) and (4.7-3), we obtain:

$$P_{ta} = \frac{\pi k e_o^2}{C} \frac{\beta_{2r}}{\beta_{2t}} \tag{4.7-4}$$

By definition, the Plank constant h is inverse proportional to the *fine structure constant* α. Since the *Universal spiral field constant U* is the inverse of the fine structure constant α ($U = 1/\alpha$), we obtain:

$$h = \frac{2\pi k e_o^2 U}{C}$$

(4.7-5)

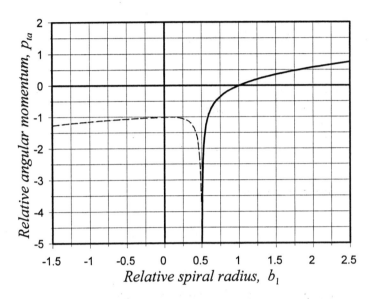

Fig. 4.7-1 Torix relative angular momentum p_{ta} as a function of relative spiral radius b_1 - Eq. (4.7-8).

Then from Eqs. (4.7-4) and (4.7-5), we find:

$$P_{ta} = \frac{h}{2U} \frac{\beta_{2r}}{\beta_{2t}}$$

(4.7-6)

From Eqs. (1.2-11), (4.1-14), (4.1-15) and (4.7-6), torix relative angular momentum p_{ta} is equal to:

$$p_{ta} = P_{ta} \frac{2U}{h} = \frac{\beta_{2r}}{\beta_{2t}} = \frac{\gamma_{2r}}{\gamma_{2t}} = \gamma_2$$

(4.7-7)

From Eqs. (4.7-7), (4.1-14) and (4.1-15), we can express torix relative angular momentum p_{ta} as a function of relative spiral radius b_1:

$$P_{ta} = \frac{b_1 - 1}{\sqrt{2b_1 - 1}} \qquad (4.7\text{-}8)$$

Fig. 4.7-1 shows a plot of Eq. (4.7-8).

4.8 Magnetic moment

Torix magnetic moment is given by:

$$\mu_t = I_t A_t \qquad (4.8\text{-}1)$$

Torix current I_t is equal to:

$$I_t = \frac{e_t V_{2t}}{2 \pi r_1} \qquad (4.8\text{-}2)$$

Torix effective cross-sectional area A_t is equal to:

$$A_t = 2 \pi r_1 (r_1 - r_i) \qquad (4.8\text{-}3)$$

From Eqs. (4.8-1) - (4.8-3), we obtain:

$$\mu_t = e_t V_{2t} (r_1 - r_i) \qquad (4.8\text{-}4)$$

Considering Eqs. (1.1-4), (4.1-10), (4.1-14), (4.4-3), (4.5-6), and (4.8-4), we obtain:

$$\mu_t = \pm \left(\frac{b_1 - 1}{2 b_1} \right)^2 \sqrt{2b_1 - 1} \, \frac{k e_o^3}{m_o C} \qquad (4.8\text{-}5)$$

Torix relative magnetic moment M_t is equal to:

$$M_t = \mu_t \frac{m_o C}{k e_o^3} = \pm \left(\frac{b_1 - 1}{2 b_1} \right)^2 \sqrt{2 b_1 - 1} \qquad (4.8\text{-}6)$$

Fig. 4.8-1 shows a plot of Eq. (4.8-6).

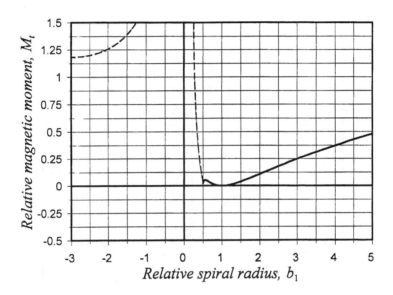

Fig. 4.8-1 Torix relative magnetic moment M_t as a function of relative spiral radius b_1 - Eq. (4.8-6).

4.9 Spiral Energy

Torix spiral energy is given by:

$$E_t = \frac{1}{2U} f_2 h \qquad (4.9\text{-}1)$$

From Eqs. (4.2-5), (4.5-7), (4.7-5), and (4.9-1), we obtain:

$$E_t = \frac{m_o C^2}{b_1} \qquad (4.9\text{-}2)$$

Torix relative spiral energy ϵ_t is equal to:

$$\epsilon_t = \frac{E_t}{m_o C^2} = \frac{1}{b_1} \qquad (4.9\text{-}3)$$

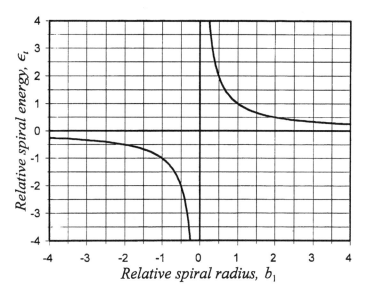

Fig. 4.9-1 Torix relative spiral energy E_t as a function
of relative spiral radius b_1 - Eq. (4.9-3).

Fig. 4.9-1 shows a plot of Eq. (4.9-3).

CHAPTER 5

PHYSICAL PROPERTIES OF HELICES

5.1 Velocities

The spiral velocity of the helix leading spiral V_3 is equal to the geometrical sum of translational velocity V_{3t} and rotational velocity V_{3r} (Figs. 5.1-1 and 5.1-2):

$$V_{3t}^2 + V_{3r}^2 = V_3^2 \qquad (5.1-1)$$

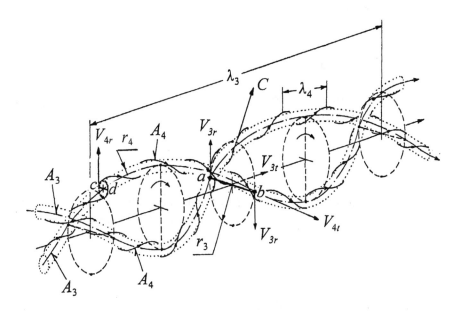

Fig. 5.1-1 Helix velocities.

Similarly, in the helix trailing spiral, the spiral velocity V_4 is equal to the geometrical sum of translational velocity V_{4t} and rotational velocity V_{4r}:

$$V_{4t}^2 + V_{4r}^2 = V_4^2 \qquad (5.1-2)$$

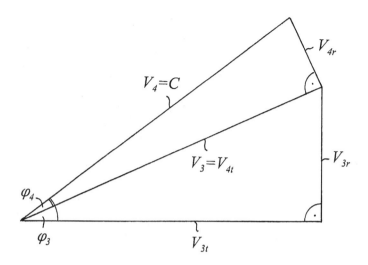

Fig. 5.1-2 Godograph of the helix velocities.

The condition of the compatibility of the velocities of the leading and trailing spirals is given by:

$$V_3 = V_{4t} \tag{5.1-3}$$

The trailing spiral velocity V_4 is greater than the leading spiral velocity V_3 and is assumed to be equal to ultimate spiral field velocity C:

$$V_4 = C \tag{5.1-4}$$

From Eqs. (5.1-1) - (5.1-4), we obtain:

$$V_{3t}^2 + V_{3r}^2 = V_{4t}^2 \tag{5.1-5}$$

$$V_{4t}^2 + V_{4r}^2 = C^2 \tag{5.1-6}$$

Later, we will use helix relative velocities β_{3t}, β_{3r}, β_{4t}, and β_{4r}. From Eqs. (3.3-2), (3.3-3), (3.3-5), and (3.3-6) and Fig. 5.1-1, we obtain:

$$\beta_{4t} = \frac{V_{4t}}{C} = \gamma_{4t} = \cos\varphi_4 \tag{5.1-7}$$

$$\beta_{4r} = \frac{V_{4r}}{C} = \gamma_{4r} = \sin\varphi_4 \tag{5.1-8}$$

$$\beta_{3t} = \frac{V_{3t}}{C} = \beta_{4t}\cos\varphi_3 = \gamma_{3t}\gamma_{4t} \tag{5.1-9}$$

$$\beta_{3r} = \frac{V_{3r}}{C} = \beta_{4t}\sin\varphi_3 = \gamma_{3r}\gamma_{4t} \tag{5.1-10}$$

Below, relative velocities β_{3t}, β_{3r}, β_{4t}, and β_{4t} are expressed as a function of spiral relative radius b_3. From Eqs. (3.3-11) and (5.1-7), we obtain:

$$\beta_{4t} = \frac{1}{b_3^{\,2}}\sqrt{b_3^{\,4} - (b_3 - 1)^4} = \cos\varphi_4 \tag{5.1-11}$$

From Eqs. (3.3-12) and (5.1-8), we obtain:

$$\beta_{4r} = \left(\frac{b_3 - 1}{b_3}\right)^2 = \sin\varphi_4 \tag{5.1-12}$$

From Eqs. (3.3-9), (3.3-11), and (5.1-9), we obtain:

$$\beta_{3t} = \frac{\sqrt{2b_3 - 1}}{b_3^{\,3}}\sqrt{b_3^{\,4} - (b_3 - 1)^4} \tag{5.1-13}$$

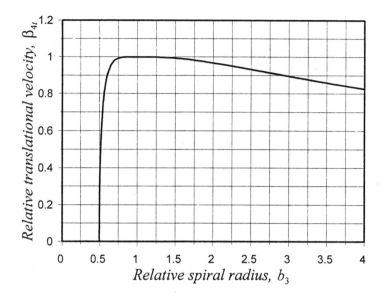

Fig. 5.1-3 Relative translational velocity β_{4t} as a function of
relative spiral radius b_3 - Eq. (5.1-11).

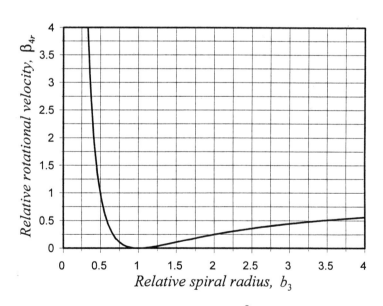

Fig. 5.1-4 Relative rotational velocity β_{4r} as a function of
relative spiral radius b_3 - Eq. (5.1-12).

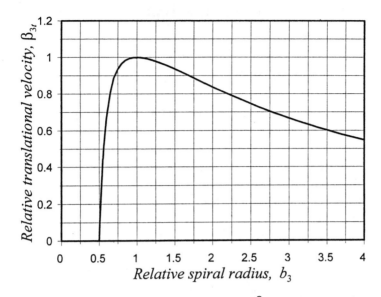

Fig. 5.1-5 Relative translational velocity β_{3t} as a function of relative spiral radius b_3 - Eq. (5.1-13).

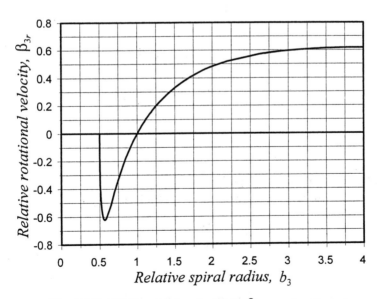

Fig. 5.1-6 Relative rotational velocity β_{3r} as a function of relative spiral radius b_3 - Eq. (5.1-14).

From Eqs. (3.1-7), (3.3-10), (3.3-11), and (5.1-10), we obtain:

$$\beta_{3r} = \frac{b_3 - 1}{b_3^{\ 3}} \sqrt{b_3^{\ 4} - (b_3 - 1)^4} = x_h \sqrt{1 - x_h^4} \qquad (5.1\text{-}14)$$

Figs. 5.1-3 - 5.1-6 show plots of Eqs. (5.1-11) - (5.1-14). Note the following features of these plots:

Fig. 5.1-3: The maximum value of relative translational velocity $\beta_{4t} = 1$ is reached at $b_3 = 1$
Fig. 5.1-4: The minimum value of relative rotational velocity $\beta_{4r} = 0$ is reached at $b_3 = 1$.
Fig. 5.1-5: The maximum value of relative translational velocity $\beta_{3t} = 1$ is reached at $b_3 = 1$
Fig. 5.1-6: The maximum negative value of relative rotational velocity β_{3r} is reached at $b_3 = 0.568$.

From Eqs. (5.1-5) - (5.1-10):

$$\beta_{3t}^2 + \beta_{3r}^2 = \beta_{4t}^2 \qquad (5.1\text{-}15)$$

$$\beta_{4t}^2 + \beta_{4r}^2 = 1 \qquad (5.1\text{-}16)$$

5.2 Spiral frequencies and densities

The frequency of helix leading spiral f_3 is equal to:

$$f_3 = \frac{V_{3r}}{2\pi r_3} \qquad (5.2\text{-}1)$$

From Eqs. (3.1-4), (4.2-1), (5.1-10), (5.2-1), we find the relative frequency of the helix leading spiral δ_3:

$$\delta_3 = \frac{f_3}{f_o} = \frac{\beta_{3r}}{b_3} \qquad (5.2-2)$$

From Eqs. (5.1-14) and (5.2-2), we can express the relative frequency of the helix leading spiral δ_3 as a function of relative spiral radius b_3:

$$\delta_3 = \frac{f_3}{f_o} = \frac{b_3 - 1}{b_3^{\,4}} \sqrt{b_3^{\,4} - (b_3 - 1)^4} \qquad (5.2-3)$$

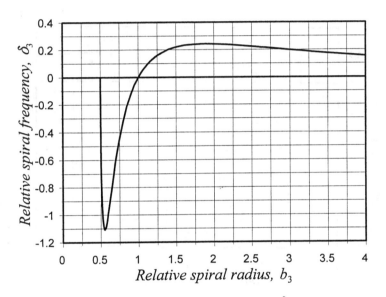

Fig. 5.2-1 Relative frequency of the helix leading spiral δ_3 as a function of the relative spiral radius b_3 - Eq. (5.2-3).

The frequency of the helix trailing spiral f_4 is equal to:

$$f_4 = \frac{V_{4r}}{2\pi r_4} \qquad (5.2-4)$$

From Eqs. (3.1-5), (3.1-6), (4.2-1), (5.1-8), and (5.2-4) we find the relative frequency of the helix trailing spiral δ_4:

$$\delta_4 - \frac{f_4}{f_o} - \frac{\beta_{4r}}{b_4} - \frac{\beta_{4r}}{b_3 - 1} \qquad (5.2\text{-}5)$$

From Eqs. (3.1-6), (5.1-12), and (5.2-5), we can express the relative frequency of the helix trailing spiral δ_4 as a function of relative spiral radius b_3:

$$\delta_4 = \frac{f_4}{f_o} = \frac{b_3 - 1}{b_3^{\,2}} \qquad (5.2\text{-}6)$$

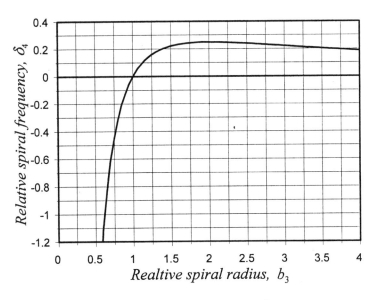

Fig. 5.2-2 Relative frequency of helix trailing spiral δ_4 as a function of relative spiral radius b_3 - Eq. (5.2-6).

Figs. 5.2-1 and 5.2-2 show plots of Eqs. (5.2-3), and (5.2-6). Note that Eq. (5.2-3) yields two values for relative spiral radius b_3 at which the relative frequency of the helix leading spiral δ_3 has the same value. The smaller value of b_3 relates to the fast helix while the greater value of b_3 relates to the slow helix. Other features of the plots are:

Fig. 5.2-1: The maximum negative value of the relative frequency of the helix leading spiral δ_3 is reached at $b_3 = 0.554$, while the maximum positive value of the relative frequency of the helix leading spiral δ_3 is reached at $b_3 = 1.90$.

Fig. 5.2-2: The maximum value of the relative frequency of the helix trailing spiral δ_4 is reached at $b_3 = 2.0$.

By definition, the density of helix leading spiral v_3 is equal to:

$$v_3 = \frac{f_3}{V_{3t}} \qquad (5.2\text{-}7)$$

From Eqs. (5.1-9), (5.1-13), (5.2-3), and (5.2-7), we can express the density of helix leading spiral v_3 as a function of relative spiral radius b_3:

$$v_3 = \frac{f_o}{C} \frac{b_3 - 1}{b_3 \sqrt{2b_3 - 1}} \qquad (5.2\text{-}8)$$

By definition, the density of the helix trailing spiral v_4 is equal to:

$$v_4 = \frac{f_4}{V_{4t}} \qquad (5.2\text{-}9)$$

From Eqs. (5.1-7), (5.1-11), (5.1-12), (5.2-5), and (5.2-9), we can express the density of helix trailing spiral v_4 as a function of relative spiral radius b_3:

$$v_4 = \frac{f_o}{C} \frac{b_3 - 1}{\sqrt{b_3^4 - (b_3 - 1)^4}} \qquad (5.2\text{-}10)$$

The relative densities of the helix leading and trailing spirals ρ_3 and ρ_4 are respectively equal to:

$$\rho_3 = v_3 \frac{C}{f_o} = \frac{b_3 - 1}{b_3\sqrt{2b_3 - 1}}$$ (5.2-11)

$$\rho_4 = v_4 \frac{C}{f_o} = \frac{b_4 - 1}{\sqrt{b_3^4 - (b_3 - 1)^4}}$$ (5.2-12)

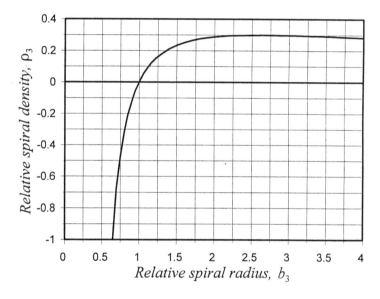

Fig. 5.2-3 Relative density of helix leading spiral ρ_3 as a function of relative spiral radius b_3 - Eq. (5.2-11).

Figs. 5.2-3 and 5.2-4 show plots of Eqs. (5.2-11), and (5.2-12). Note the following features of these plots:

Fig. 5.2-3: The maximum value of the relative density of helix leading spiral ρ_3 is reached at $b_3 = 2.618$
Fig. 5.2-4: The maximum value of the relative density of helix trailing spiral ρ_4 is reached at $b_3 = 2.191$.

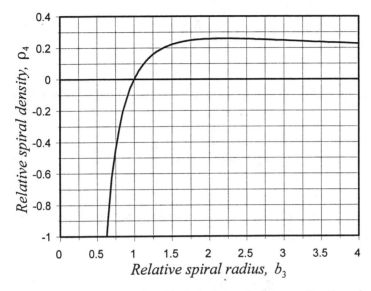

Fig. 5.2-4 Relative density of helix trailing spiral ρ_4 as a function of relative spiral radius b_3 - Eq. (5.2-12).

5.3 Helix spiral energy

Photon energy is proportional to the product of the helix frequency f_3 and the Plank constant h. For reality-polarized helicies, the real and imaginary spiral energies E_h and $\bar{E}_h$ of the helix are respectively equal to:

$$E_h = \frac{f_3 h}{4} \tag{5.3-1}$$

$$\bar{E}_h = -\frac{f_3 h}{4} \tag{5.3-2}$$

From Eqs. (5.2-2) and (5.3-1), we obtain:

$$E_h = \frac{\delta_3 f_o h}{4}$$ (5.3-3)

From Eqs. (4.5-7), (4.7-5), and (5.3-3), helix real spiral energy E_h is equal to:

$$E_h = abs\left(Um_o C^2 \frac{\delta_3}{2} \right)$$ (5.3-4)

Similarly, from Eqs. (4.5-7), (4.7-5), and (5.3-2), helix imaginary spiral energy $\overline{E}_h$ is equal to:

$$\overline{E}_h = -abs\left(Um_o C^2 \frac{\delta_3}{2} \right)$$ (5.3-5)

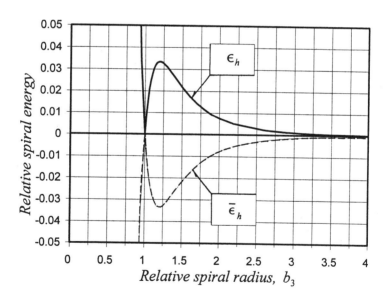

Fig. 5.3-1 Helix real and imaginary relative spiral energies ϵ_h and $\overline{\epsilon}_h$ as a function of relative spiral radius b_3 - Eqs. (5.3-6) and (5.3-7).

From Eqs. (5.2-3), (5.3-4), and (5.3-5), we find the real and imaginary relative spiral energies ϵ_h and $\bar{\epsilon}_h$ of the helix (Fig. 5.3-1):

$$\epsilon_h = \frac{E_h}{Um_oC^2} = abs\left(\frac{b_3 - 1}{2b_3^4}\sqrt{b_3^4 - (b_3 - 1)^4}\right) \qquad (5.3\text{-}6)$$

$$\bar{\epsilon}_h = \frac{\bar{E}_h}{Um_oC^2} = -abs\left(\frac{b_3 - 1}{2b_3^4}\sqrt{b_3^4 - (b_3 - 1)^4}\right) \qquad (5.3\text{-}7)$$

5.4 Helix electric charge and gravitational mass

From assumption and Eqs. (5.1-14), relative helix electric charge e_h/e_o is equal to:

$$\frac{e_h}{e_o} = \frac{\beta_{3r}}{4} = \pm abs\left(\frac{x_h}{4}\sqrt{1 - x_h^4}\right) \qquad (5.4\text{-}1)$$

From Eqs. (5.1-14) and (5.4-1), we can express helix relative electric charge e_h/e_o as a function of the relative spiral radius b_3:

$$\frac{e_h}{e_o} = \pm abs\left(\frac{b_3 - 1}{4b_3^3}\sqrt{b_3^4 - (b_3 - 1)^4}\right) \qquad (5.4\text{-}2)$$

Fig. 5.4-1 shows a plot of Eqs. (5.4-2).

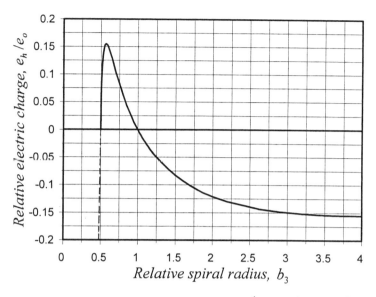

Fig. 5.4-1 Helix relative electric charge e_h/e_o as a function of relative spiral radius b_3 - Eq. (5.4-2).

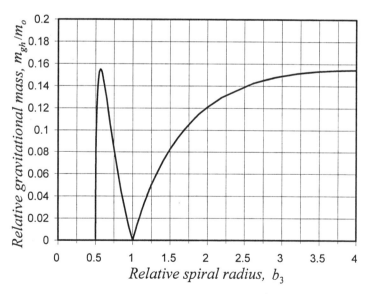

Fig. 5.4-2 Helix relative gravitational mass m_{hg}/m_o as a function of relative spiral radius b_3 - Eq. (5.4-4).

Considering Eq. (5.1-14), the helix relative gravitational mass m_{hg}/m_o is assumed to be equal to:

$$\frac{m_{hg}}{m_o} = abs\left(\frac{\beta_{3r}}{4}\right) = abs\left(\frac{x_h}{4}\sqrt{1 - x_h^4}\right) \tag{5.4-3}$$

From Eqs. (5.1-14) and (5.4-3), we can express the helix relative gravitational mass m_{hg}/m_o as a function of the relative spiral radius b_3:

$$\frac{m_{hg}}{m_o} = abs\left(\frac{b_3 - 1}{4b_3^3}\sqrt{b_3^4 - (b_3 - 1)^4}\right) \tag{5.4-4}$$

Fig. 5.4-2 shows a plot of Eqs. (5.4-4).

5.5 Fast and slow helices

Eq. (5.2-3) yields two values of the relative spiral radius b_{3f} and b_{3s} corresponding to the same relative spiral frequency δ_3:

$$\delta_3 = \frac{b_{3f} - 1}{b_{3f}^4}\sqrt{b_{3f}^4 - (b_{3f} - 1)^4} = \frac{b_{3s} - 1}{b_{3s}^4}\sqrt{b_{3s}^4 - (b_{3s} - 1)^4} \tag{5.5-1}$$

Subsequently, Eq. (5.1-13) yields two respective values for relative translational velocity, fast β_{3tf} and slow β_{3ts}, corresponding to the same relative spiral frequency δ_3. Helices with faster translational velocity are called *fast helices*, while helices with slower translational velocity are called *slow helices*. Equations below establish approximate relationships between relative spiral frequency δ_3 and relative spiral radii b_{3f} and b_{3s} of fast and slow helices respectively.

For fast helices, $b_{3f} \sim 1$. Therefore, we obtain from Eq. (5.5-1):

$$\delta_3 \approx b_{3f} - 1 \tag{5.5-2}$$

$$b_{3f} \approx 1 + \delta_3 \tag{5.5-3}$$

For the slow helices, $b_{3s} \gg 1$. Therefore, we obtain from Eq. (5.5-1):

$$\delta_3 \approx 2 \left(b_{3s} \right)^{-3/2} \qquad (5.5\text{-}4)$$

CHAPTER 6

CLASSIFICATION OF TORICES

6.1 Quantum change of torix geometry

Torix geometry changes in quantum steps. Quantum change of torix geometry occurs either by oscillation or by excitation (Fig. 6.1-1). During the oscillation, all the torix dimensions change proportionally by the same factor, called the *torix oscillation factor Q*. During excitation of a torix, the spiral radius of its leading spiral r_1 changes, while its inversion radius r_i remains constant.

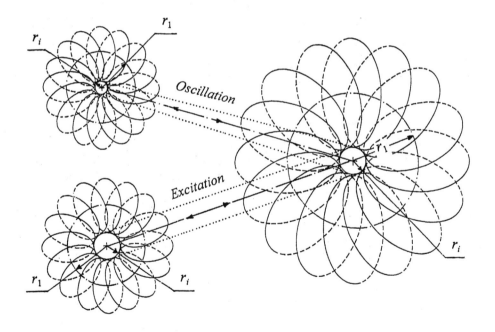

Fig. 6.1-1 Oscillation and excitation of a torix.

6.2 Oscillation of torices

It is convenient to express the torix oscillation factor Q as a ratio of the inversion radius r_{io} of the torix at oscillation level $N = 0$ to the inversion radius r_{iN} of the torix at oscillation level N as given by:

$$Q = \frac{r_{io}}{r_{iN}} \tag{6.2-1}$$

From Eqs. (4.5-6), (4.5-7), and (6.2-1), we find oscillation factor Q to be equal to the two ratios: (1) ratio of the rest mass of reality polarized torices m_{oN} at oscillation level N to the rest mass of reality-polarized torices m_{oo} at oscillation level $N = 0$, and (2) ratio of the fundamental spiral field frequency f_{oN} of the torix at oscillation level N to the fundamental spiral field frequency f_{oo} of the torix at oscillation level $N = 0$, as given by:

$$Q = \frac{m_{oN}}{m_{oo}} = \frac{f_{oN}}{f_{oo}} \tag{6.2-2}$$

6.3 Classification of oscillation torices

Oscillation torices are divided into two groups, base and supplementary. Oscillation levels N of base oscillation torices vary from 0 to 27. The first three oscillation levels $N = 0$, 1, and 2 are called *harmonic*, while the higher oscillation levels are called *universal*.

For harmonic oscillation levels, the oscillation factor Q is given by:

$$Q = N + 1 \qquad N = 0, 1, 2. \tag{6.3-1}$$

For universal oscillation levels, the oscillation factor Q is defined by the equation:

$$Q = 3 \left[\frac{U}{2(N-2)} \right]^{N-2} \qquad N = 3, 4, 5. \ . \ . \tag{6.3-2}$$

Table 6.3-1 shows the parameters of base oscillation torices corresponding to the first four and the last oscillation level N of this group. The torix corresponding to oscillation level $N = 0$. Note that the rest electric charge of all oscillation torices is equal to the rest electric charge of electron e_{oN}.

Table 6.3-1 Parameters of base oscillation torices ($0 \leq N \leq 27$).

Oscillation levels	N	Q	m_{oN}	f_{oN}	r_{iN}
			MeV/c^2	Hz	m
Harmonic	0	1.000	0.511	3.386×10^{22}	1.409×10^{-15}
	1	2.000	1.022	6.772×10^{22}	7.045×10^{-16}
	2	3.000	1.533	1.016×10^{23}	4.697×10^{-16}
Universal	3	205.554	105.038	6.961×10^{24}	6.855×10^{-18}
	4	3521.04	1799.246	1.192×10^{26}	4.002×10^{-19}
	5	35741.4	18263.815	1.210×10^{27}	3.942×10^{-20}
	..	....	....	....	....
	27	2.653×10^{11}	1.356×10^{11}	8.984×10^{33}	5.311×10^{-27}

Table 6.3-2 Parameters of supplementary oscillation torices ($N < 28$).

Oscillation levels	N	Q	m_{oN}	f_{oN}	r_{iN}
			MeV/c^2	Hz	m
Universal	28	2.623×10^{11}	1.340×10^{11}	8.881×10^{33}	5.372×10^{-27}
	29	2.495×10^{11}	1.275×10^{11}	8.448×10^{33}	5.648×10^{-27}
	30	2.287×10^{11}	1.169×10^{11}	7.744×10^{33}	6.161×10^{-27}
	..	....	....	....	....
	98	2.608×10^{-24}	1.333×10^{-14}	8.832×10^{08}	5.402×10^{-02}
	99	6.813×10^{-25}	3.481×10^{-15}	2.307×10^{08}	2.068×10^{-01}
	100	1.761×10^{-25}	9.000×10^{-16}	5.964×10^{07}	8.000×10^{-01}

Oscillation levels N of supplementary oscillation torices vary from 28 to $+\infty$.   Table 6.3-2 shows the parameters of supplementary oscillation torices corresponding to several oscillation levels N of this group. Torix parameters are defined in Tables 6.3-1 and 6.3-2 from Eqs. (4.5-6), (4.5-7), (6.2-1), (6.2-2), (6.3-1), and (6.3-2).

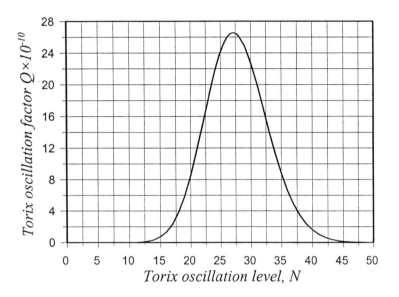

Fig. 6.3-1 Torix oscillation factor Q as a function
of oscillation level N - Eq. (6.3-2).

Fig. 6.3-1 shows a plot of Eq. (6.3-2). The maximum value of oscillation factor Q corresponds to oscillation level N_m that is equal to:

$$N_m = 1 + \frac{U}{2e} \quad (e = 2.718) \tag{6.3-3}$$

6.4 Groups of torices

Torices are divided into groups that are identified by group number g shown in Table 6.4-1 and Fig. 6.4-1. Torices of each group are located within the ranges of relative spiral radius b_1 that are given by the relationship:

$$\frac{1}{g+1} \leq b_1 \leq \frac{1}{g} \qquad g = 0, \pm 1, \pm 2, \pm 3, \ldots \qquad (6.4\text{-}1)$$

Table 6.4-1 Several group ranges of torix relative spiral radius b_1.

Positive group				Negative group			
g	Symbol	Range of b_1		g	Symbol	Range of b_1	
		From	To			From	To
0	$\pm A$	2	$+\infty$	-1	$\pm \overline{A}$	$-\infty$	-2
	$\pm A$	1	2		$\pm \overline{A}$	-2	-1
1	$\pm a$	2/3	1	-2	$\pm \overline{B}$	-1	-2/3
	$\pm a$	1/2	2/3		$\pm \overline{B}$	-2/3	-1/2
2	$\pm \overline{a}$	2/5	1/2	-3	$\pm \overline{C}$	-1/2	-2/5
	$\pm \overline{a}$	1/3	2/5		$\pm \overline{C}$	-2/5	-1/3
3	$\pm \overline{b}$	2/7	1/3	-4	$\pm \overline{D}$	-1/3	-2/7
	$\pm \overline{b}$	1/4	2/7		$\pm \overline{D}$	-2/6	-1/4
4	$\pm \overline{c}$	2/9	1/4	-5	$\pm \overline{E}$	-1/4	-2/9
	$\pm \overline{c}$	1/5	2/9		$\pm \overline{E}$	-2/9	-1/5

Designation of torices is shown below.

A, a, B, b, etc. . . . . Outer complementary-polarized torices
A, a, B, b, etc. . . . . .Inner complementary-polarized torices
A, A, B, B . . . . .Real torices
$\overline{A}, \overline{A}, \overline{B}, \overline{B}, \overline{a}, \overline{a}, \overline{b}, \overline{b}$, etc. . . . .Imaginary torices.

The first subscript in the torix symbol represents the oscillation level N, while the second subscript represents the excitation level, either universal *n* or harmonic *m*.

6.5 Relationship between universal excitation torices

Properties of the universal excitation torices are expressed as a function of the *universal quantum parameter z*. This parameter is the product of universal level *n* and the *universal spiral field constant U* as given by:

$$z = nU \qquad n = 0, 1, 2, \ldots \qquad (6.5\text{-}1)$$

Table 6.5-1 Parameters of universal excitation torices ($b_1 > 0$).

Torix		Quantum values of b_1	Relationship with matched torix	Matched torix	
Name	g			Name	g
$\pm A_{Nn}$	0	$2(1 + z^2)$	Reality-polarized torix in respect to	$\pm \bar{A}_{Nn}$	-1
$\pm A_{Nn}$	0	$\dfrac{2(1 + z^2)}{1 + 2z^2}$	Complementary-polarized torix in respect to	$\pm A_{Nn}$	0
$\pm a_{Nn}$	1	$\dfrac{2(1 + z^2)}{3 + 2z^2}$	Vorticity-polarized torix in respect to	$\pm A_{Nn}$	0
$\pm a_{Nn}$	1	$\dfrac{2(1 + z^2)}{3 + 4z^2}$	Vorticity-polarized torix in respect to	$\pm A_{Nn}$	0
$\pm \bar{a}_{Nn}$	2	$\dfrac{2(1 + z^2)}{5 + 4z^2}$	Vorticity-polarized torix in respect to	$\pm \bar{A}_{Nn}$	-1
$\pm \bar{a}_{Nn}$	2	$\dfrac{2(1 + z^2)}{5 + 6z^2}$	Vorticity-polarized torix in respect to	$\pm \bar{A}_{Nn}$	-1
$\pm \bar{b}_{Nn}$	3	$\dfrac{2(1 + z^2)}{7 + 6z^2}$	Vorticity-polarized torix in respect to	$\pm \bar{B}_{Nn}$	-2
$\pm \bar{b}_{Nn}$	3	$\dfrac{2(1 + z^2)}{7 + 8z^2}$	Vorticity-polarized torix in respect to	$\pm \bar{B}_{Nn}$	-2

Table 6.5-1 shows the quantum values of the relative spiral radius b_1 of universal excitation torices when $b_1 > 0$, and the relationship between matched torices. Table 6.5-2 shows quantum values of relative spiral radius b_1 of the universal torices when $b_1 < 0$, and the relationship between the matched torices.

Table 6.5-2 Parameters of universal excitation torices $(b_1 < 0)$.

Torix		Quantum levels of b_1	Relationship with matched torix	Matched torix	
Name	g			Name	g
$\pm \bar{A}_{Nn}$	-1	$-2(1 + z^2)$	Reality-polarized torix in respect to	$\pm A_{Nn}$	0
$\pm \bar{A}_{Nn}$	-1	$-\dfrac{2(1 + z^2)}{1 + 2z^2}$	Reality-polarized torix in respect to	$\pm A_{Nn}$	0
$\pm \bar{B}_{Nn}$	-2	$-\dfrac{2(1 + z^2)}{3 + 2z^2}$	Reality and vorticity-polarized torix in respect to	$\pm a_{Nn}$	1
$\pm \bar{B}_{Nn}$	-2	$-\dfrac{2(1 + z^2)}{3 + 4z^2}$	Reality and vorticity-polarized torix in respect to	$\pm a_{Nn}$	1
$\pm \bar{C}_{Nn}$	-3	$-\dfrac{2(1 + z^2)}{5 + 4z^2}$	Reality and vorticity-polarized torix in respect to	$\pm \bar{a}_{Nn}$	2
$\pm \bar{C}_{Nn}$	-3	$-\dfrac{2(1 + z^2)}{5 + 6z^2}$	Reality and vorticity-polarized torix in respect to	$\pm \bar{a}_{Nn}$	2
$\pm \bar{D}_{Nn}$	-4	$-\dfrac{2(1 + z^2)}{7 + 6z^2}$	Reality and vorticity-polarized torix in respect to	$\pm \bar{b}_{Nn}$	3
$\pm \bar{D}_{Nn}$	-4	$-\dfrac{2(1 + z^2)}{7 + 8z^2}$	Reality and vorticity-polarized torix in respect to	$\pm \bar{b}_{Nn}$	3

6.6 Relationship between harmonic excitation torices

Properties of harmonic excitation torices are expressed as a function of the parameter $m = 0, 1, 2, \ldots$. Table 6.6-1 shows the quantum values of relative spiral radius b_1 of harmonic excitation torices when $b_1 > 0$, and the relationships between matched torices. Table 6.6-2 shows the quantum values of the relative spiral radius b_1 of the harmonic torices with $b_1 < 0$, and the relationship between the matched torices.

Table 6.6-1 Parameters of harmonic excitation torices ($b_1 > 0$).

Torix		Quantum values of b_1	Relationship with matched torix	Matched torix	
Name	g			Name	g
$\pm A_{Nm}$	0	$2 + m$	Reality-polarized torix in respect to	$\pm \overline{A}_{Nm}$	-1
$\pm A_{Nm}$	0	$\dfrac{2 + m}{1 + m}$	Complementary-polarized torix in respect to	$\pm A_{Nm}$	0
$\pm a_{Nm}$	1	$\dfrac{2 + m}{3 + m}$	Vorticity-polarized torix in respect to	$\pm A_{Nm}$	0
$\pm a_{Nm}$	1	$\dfrac{2 + m}{3 + 2m}$	Vorticity-polarized torix in respect to	$\pm A_{Nm}$	0
$\pm \overline{a}_{Nm}$	2	$\dfrac{2 + m}{5 + 2m}$	Vorticity-polarized torix in respect to	$\pm \overline{A}_{Nm}$	-1
$\pm \overline{a}_{Nm}$	2	$\dfrac{2 + m}{5 + 3m}$	Vorticity-polarized torix in respect to	$\pm \overline{A}_{Nm}$	-1
$\pm \overline{b}_{Nm}$	3	$\dfrac{2 + m}{7 + 3m}$	Vorticity-polarized torix in respect to	$\pm \overline{B}_{Nm}$	-2
$\pm \overline{b}_{Nm}$	3	$\dfrac{2 + m}{7 + 4m}$	Vorticity-polarized torix in respect to	$\pm \overline{B}_{Nm}$	-2

Table 6.6-2 Parameters of harmonic excitation torices ($b_1 < 0$).

Torix		Quantum levels of b_1	Relationship with matched torix	Matched torix	
Name	g			Name	g
$\pm \bar{A}_{Nm}$	-1	$-(2+m)$	Reality-polarized torix in respect to	$\pm A_{Nm}$	0
$\pm \bar{A}_{Nm}$	-1	$-\dfrac{2+m}{1+m}$	Reality-polarized torix in respect to	$\pm A_{Nm}$	0
$\pm \bar{B}_{Nm}$	-2	$-\dfrac{2+m}{3+m}$	Reality and vorticity-polarized torix in respect to	$\pm a_{Nm}$	1
$\pm \bar{B}_{Nm}$	-2	$-\dfrac{2+m}{3+2m}$	Reality and vorticity-polarized torix in respect to	$\pm a_{Nm}$	1
$\pm \bar{C}_{Nm}$	-3	$-\dfrac{2+m}{5+2m}$	Reality and vorticity-polarized torix in respect to	$\pm \bar{a}_{Nm}$	2
$\pm \bar{C}_{Nm}$	-3	$-\dfrac{2+m}{5+3m}$	Reality and vorticity-polarized torix in respect to	$\pm \bar{a}_{Nm}$	2
$\pm \bar{D}_{Nm}$	-4	$-\dfrac{2+m}{7+3m}$	Reality and vorticity-polarized torix in respect to	$\pm \bar{b}_{Nm}$	3
$\pm \bar{D}_{Nm}$	-4	$-\dfrac{2+m}{7+4m}$	Reality and vorticity-polarized torix in respect to	$\pm \bar{b}_{Nm}$	3

6.7 Relationship between matched torices

Relationship between three types of matched torices are considered below:

- Reality-polarized torices
- Vorticity-polarized torices
- Complementary-polarized torices.

Reality-polarized torices - From Eqs. (2.2-2), (2.2-4), and (4.2-5), we find the sums of the relative spiral frequencies of real torices δ_2 and imaginary torices $\bar{\delta}_2$.

For a negative torix, we obtain:

$$\delta_2 + \bar{\delta}_2 = 0 \tag{6.7-1}$$

For a positive torix, we obtain:

$$\delta_2 + \bar{\delta}_2 = 4 \tag{6.7-2}$$

From Eqs. (2.2-2) and (4.4-3), we find the sums of relative electric charges of real and imaginary torices.

For negative torices, we obtain:

$$\frac{e_t}{e_o} + \frac{\bar{e}_t}{e_o} = -1 \tag{6.7-3}$$

For positive torices, we obtain:

$$\frac{e_t}{e_o} + \frac{\bar{e}_t}{e_o} = 1 \tag{6.7-4}$$

From Eqs. (2.2-2) and (4.3-3), we find that for both the negative and positive reality-polarized torices the sum of relative gravitational masses of real and imaginary torices is equal to:

$$\frac{m_{tg}}{m_o} + \frac{\bar{m}_{tg}}{m_o} = 1 \tag{6.7-5}$$

Vorticity-polarized torices - From Eqs. (2.3-3) and (4.2-5), we find that for these torices, the sum of relative spiral frequencies of positive

δ_2^+ and negative δ_2^- torices is equal to:

$$\delta_2^+ + \delta_2^- = 2 \qquad (6.7\text{-}6)$$

From Eqs. (2.3-4), (4.2-5), (4.4-3), and (6.7-6), we find that for vorticity-polarized torices the sum of relative electric charges of positive and negative torices is equal to zero. Therefore, we obtain:

$$\frac{e_t^+}{e_o} = -\frac{e_t^-}{e_o} \qquad (6.7\text{-}7)$$

Complementary-polarized torices - From Eqs. (2.4-2), (2.4-8), and (4.2-5), we find the sums of relative spiral frequencies of inner δ_2' and outer δ_2'' complementary-polarized torices.
For negative torices, we obtain:

$$\delta_2' + \delta_2'' = 1 \qquad (6.7\text{-}8)$$

For positive torices, we obtain:

$$\delta_2' + \delta_2'' = 3 \qquad (6.7\text{-}9)$$

From Eqs. (2.4-3), (4.2-5), (4.4-3), (6.7-8), and (6.7-9), we find the sums of electric charges of inner and outer complementary-polarized torices.
For negative torices, we obtain:

$$\frac{e_t'}{e_o} + \frac{e_t''}{e_o} = -\frac{1}{2} \qquad (6.7\text{-}10)$$

For positive torices, we obtain:

$$\frac{e_t^{\,\prime}}{e_o} + \frac{e_t^{\,\prime\prime}}{e_o} = +\frac{1}{2} \tag{6.7-11}$$

NOMENCLATURE

A_t = Torix effective cross-section area

b_A = Relative radius of outer circular distribution of mini charges

b_a = Relative radius of inner circular distribution of mini charges

b_B = Relative radius of outer circular distribution of mini charges

b_b = Relative radius of inner circular distribution of mini charges

b_1 = Relative spiral radius of torix leading spiral

b_1^- = Relative spiral radius of leading spiral in negative vorticity-polarized torix

b_1^+ = Relative spiral radius of leading spiral in positive vorticity-polarized torix

$b_1^{/}$ = Relative spiral radius of leading spiral in inner complementary-polarized torix

$b_1^{//}$ = Relative spiral radius of leading spiral in outer complementary-polarized torix

$\bar{b}_1$ = Relative spiral radius of leading spiral in a reality-polarized torix

$\bar{b}_2$ = Relative spiral radius of trailing spiral in a reality-polarized torix

b_{1j} = Relative spiral radius of torix leading spiral related to the excitation level $n = j$ (or $m = j$)

b_{1k} = Relative spiral radius of torix leading spiral related to the excitation level $n = k$ (or $m = k$)

b_2 = Relative spiral radius of torix trailing spiral

b_2^- = Relative spiral radius of trailing spiral in negative vorticity-polarized torix

b_2^+ = Relative spiral radius of trailing spiral in positive vorticity-polarized torix

$b_2^{/}$ = Relative spiral radius of trailing spiral in inner complementary-polarized torix

239

b_2'' = Relative spiral radius of trailing spiral in outer complementary-polarized torix

b_3 = Relative spiral radius of helix leading spiral

b_{3f} = Relative spiral radius of leading spiral of fast helix

b_{3s} = Relative spiral radius of leading spiral of slow helix

b_4 = Relative spiral radius of helix trailing spiral

C = Ultimate spiral field velocity

c = Velocity of light

E_h = Spiral energy of real reality-polarized helix

$\overline{E}_h$ = Spiral energy of imaginary reality-polarized helix

E_t = Torix spiral energy

e_a = Mini electric charge at point a

e_b = Mini electric charge at point b

e_h = Helix electric charge

e_o = Electron rest electric charge, rest electric charge of reality-polarized torices

e_t = Torix electric charge

e_{tA} = Electric charge of the torix with spiral radius r_A

e_{tB} = Electric charge of the torix with spiral radius r_B

e_{tb} = Electric charge of the torix with spiral radius r_b

e_{ta} = Electric charge of the torix with spiral radius r_a

$\overline{e}_t$ = Electric charge of imaginary reality-polarized torix

e_t^+ = Electric charge of positive vorticity-polarized torix

e_t^- = Electric charge of negative vorticity-polarized torix

e_1 = Absolute value of total electric mini charges in gravitational spiral field

e_2 = Absolute value of total electric mini charges in gravitational spiral field

e/m = Electromass

$F(AB)$ = Electric force between mini electric charges located at the points A and B

$F(Ab)$ = Electric force between mini electric charges located at the points A and b

$F(aB)$ = Electric force between mini electric charges located at the points a and B

$F(ab)$ = Electric force between mini electric charges located at the points a and b

F_e = Electric force

F_g = Gravitational force

F_i = Inertial force

$f(AB)$ = Relative electric force between mini electric charges located at the points A and B

$f(Ab)$ = Relative electric force between mini electric charges located at the points A and b

$f(aB)$ = Relative electric force between mini electric charges located at the points a and B

$f(ab)$ = Relative electric force between mini electric charges located at the points a and b

f_g = Relative gravitational force

f_h = Horizontal component of the relative electric (Coulomb's) force between mini electric charges

$f_h(AB)$ = Horizontal component of relative electric force between mini electric charges located at the points A and B

$f_h(Ab)$ = Horizontal component of relative electric force between mini electric charges located at the points A and b

$f_h(aB)$ = Horizontal component of relative electric force between mini electric charges located at the points a and B

$f_h(ab)$ = Horizontal component of relative electric force between mini electric charges located at the points a and b

f_o = Fundamental spiral field frequency

f_{oN} = Fundamental spiral field frequency related to oscillation level N

f_{oo} = Fundamental spiral field frequency related to oscillation level $N = 0$

f_s = Relative strong force between nucleons

f_1 = Frequency of torix leading spiral

f_2 = Frequency of torix trailing spiral, or torix spiral frequency

f_3 = Frequency of helix leading spiral, or helix spiral frequency

f_4 = Frequency of helix trailing spiral

G = Gravitational constant

g = Torix group number

h = Plank constant

I_t = Torix current

k = Coulomb constant

L = Length of one winding of a spiral

L_1 = Length of one winding of torix leading spiral

L_2 = Length of one winding of torix trailing spiral

L_3 = Length of one winding of helix leading spiral

L_4 = Length of one winding of helix trailing spiral

M_t = Torix relative magnetic moment

m = Torix harmonic excitation level

m_{hg} = Helix gravitational mass

m_i = Particle inertial mass

m_o = Rest mass of electron, rest mass of reality-polarized torices

m_{ob} = Base particle rest mass

m_{oj} = Torix rest mass related to the oscillation level $N = j$

m_{ok} = Torix rest mass related to the oscillation level $N = k$

m_{oN} = Particle rest mass related to oscillation level N, rest mass of reality-polarized torices related to oscillation level N

m_{oo} = Particle rest mass related to oscillation level $N = 0$, rest mass of reality-polarized torices related to oscillation level $N = 0$

m_{op} = Rest mass of proton

$m_{o\mu}$ = Rest mass of muon

$m_{o\tau}$ = Rest mass of tau

m_{tg} = Torix gravitational mass

$\overline{m}_{tg}$ = Gravitational mass of imaginary reality-polarized torix

m_{ti} = Torix inertial mass

m_1 = Gravitational mass of body 1

m_2 = Gravitational mass of body 2

N = Torix oscillation level

N_j = Torix universal excitation level related to the oscillation level $N = j$

N_k = Torix universal excitation level related to the oscillation level $N = k$

N_m = Torix oscillation level at maximum value of Q

n = Torix universal excitation level

n_e = A number of circularly distributed electric mini charges

n_j = Torix universal excitation level $n = j$

n_k = Torix universal excitation level $n = k$

P_{ta} = Torix angular momentum

p_{ta} = Torix relative angular momentum

Q = Torix oscillation factor

r_A = Radius of outer circular distribution of mini charges

r_a = Radius of inner circular distribution of mini charges

r_B = Radius of outer circular distribution of mini charges

r_b = Radius of inner circular distribution of mini charges

r_i = Inversion radius of spiral field

r_{iN} = Inversion radius of spiral field related to oscillation level N

r_{io} = Inversion radius related to oscillation level $N = 0$

r_j = Inversion radius of gravitational spiral field

r_1 = Spiral radius of torix leading spiral

r_2 = Spiral radius of torix trailing spiral

r_3 = Spiral radius of helix leading spiral

r_4 = Spiral radius of helix trailing spiral

S = Distance between nucleons or bodies

s_c = Relative critical distance between nucleons

U = Universal spiral field constant

V_1 = Spiral velocity of torix leading spiral

V_{1r} = Rotational velocity of torix leading spiral

V_{1t} = Translational velocity of torix leading spiral

V_2 = Spiral velocity of torix trailing spiral

V_{2r} = Rotational velocity of torix trailing spiral

V_{2t} = Translational velocity of torix trailing spiral

V_3 = Spiral velocity of helix leading spiral

V_{3r} = Rotational velocity of helix leading spiral

V_{3t} = Translational velocity of helix leading spiral

V_4 = Spiral velocity of helix trailing spiral

V_{4r} = Rotational velocity of helix trailing spiral

V_{4t} = Translational velocity of helix trailing spiral

w_1 = A number of windings in torix leading spiral

w_2 = A number of windings in torix trailing spiral

w_3 = A number of windings in helix leading spiral

w_4 = A number of windings in helix trailing spiral

x_h = Helix vortex ratio

x_t = Torix vortex ratio

z = Universal quantum parameter

α = Fine structure constant

α_j = Angular position of the mini charges distributed along the circle with radius r_A and r_a

β_k = Angular position of the mini charges distributed along the circle with radius r_B and r_b

β_{1r} = Relative rotational velocity of torix leading spiral

β_{1t} = Relative translational velocity of torix leading spiral

β_{2r} = Relative rotational velocity of torix trailing spiral

β_{2t} = Relative translational velocity of torix trailing spiral

β_{3r} = Relative rotational velocity of helix leading spiral

β_{3t} = Relative translational velocity of helix leading spiral

β_{3tf} = Relative translational velocity of leading spiral of fast helix

β_{3ts} = Relative translational velocity of leading spiral of slow helix

β_{4r} = Relative rotational velocity of helix trailing spiral

β_{4t} = Relative translational velocity of helix trailing spiral

γ_1 = Spiral slope of torix leading spiral

γ_{1r} = Rotational spirality of torix leading spiral

γ_{1t} = Translational spirality of torix leading spiral

γ_2 = Slope of torix trailing spiral

γ_{2r} = Rotational spirality of torix trailing spiral

$\overline{\gamma}_{2r}$ = Rotational spirality of trailing spiral in reality-polarized torix

γ_{2r}^{+} = Rotational spirality of trailing spiral in positive vorticity-polarized torix

γ_{2r}^{-} = Rotational spirality of trailing spiral in negative vorticity-polarized torix

$\gamma_{2r}^{'}$ = Rotational spirality of inner trailing spiral in negative complementary-polarized torix

$\gamma_{2r}^{''}$ = Rotational spirality of outer trailing spiral in negative complementary-polarized torix

γ_{2t} = Translational spirality of torix trailing spiral

γ_{2t}^{+} = Translational spirality of trailing spiral in positive vorticity-polarized torix

γ_{2t}^{-} = Translational spirality of trailing spiral in negative vorticity-polarized torix

γ_3 = Slope of helix leading spiral

γ_{3r} = Rotational spirality of helix leading spiral

γ_{3t} = Translational spirality of helix leading spiral

γ_4 = Spiral slope of helix trailing spiral

γ_{4r} = Rotational spirality of helix trailing spiral

γ_{4t} = Translational spirality of helix trailing spiral

δ_1 = Relative spiral frequency of torix leading spiral

δ_2 = Relative spiral frequency of torix trailing spiral, or torix relative spiral frequency

δ_2' = Relative spiral frequency of inner complementary-polarized torix

δ_2'' = Relative spiral frequency of outer complementary-polarized torix

$\bar{\delta}_2$ = Relative spiral frequency of imaginary reality-polarized torix

δ_2^+ = Relative spiral frequency of positive vorticity-polarized torix

δ_2^- = Relative spiral frequency of negative vorticity-polarized torix

δ_3 = Relative spiral frequency of helix leading spiral, or helix relative spiral frequency

δ_4 = Relative spiral frequency of helix trailing spiral

ϵ_h = Relative spiral energy of real reality-polarized helix

$\bar{\epsilon}_h$ = Relative spiral energy of imaginary reality-polarized helix

ϵ_t = Torix relative spiral energy

η_1 = Relative wavelength of torix leading spiral

η_2 = Relative wavelength of torix trailing spiral

η_3 = Relative wavelength of helix leading spiral

η_4 = Relative wavelength of helix trailing spiral

λ = Spiral wavelength

λ_1 = Wavelength of torix leading spiral

λ_2 = Wavelength of torix trailing spiral

λ_3 = Wavelength of helix leading spiral

λ_4 = Wavelength of helix trailing spiral

μ = Particle magnetic moment

μ_B = Bohr magneton

μ_b = Base magneton

μ_N = Nuclear magneton

μ_n = Neutron magnetic moment

μ_o = Electron-based magneton

μ_p = Proton magnetic moment

μ_t = Torix magnetic moment

v_1 = Spiral density of torix leading spiral

v_2 = Spiral density of torix trailing spiral

v_3 = Spiral density of helix leading spiral

v_4 = Spiral density of helix trailing spiral

ρ_1 = Relative spiral density of torix leading spiral

ρ_2 = Relative spiral density of torix trailing spiral

ρ_3 = Relative spiral density of helix leading spiral

ρ_4 = Relative spiral density of helix trailing spiral

φ_1 = Slope angle of torix leading spiral

φ_2 = Slope angle of torix trailing spiral

φ_3 = Slope angle of helix leading spiral

φ_4 = Slope angle of helix trailing spiral

$\varkappa$ = Electric mini charge distribution factor

Ω = Gravitational electromass ratio.

APPENDIX A

PHYSICAL CONSTANTS

c	= Velocity of light[1]	$2.997\ 924\ 58 \times 10^8$ m/s
C	= Ultimate spiral field velocity[*]	$(1.000000000000003) \times c$
h	= Planck constant[1]	$6.626\ 075\ (40) \times 10^{-34}$ J·s
e_o	= Electron rest electric charge[1]	$1.602\ 177\ 33\ (49) \times 10^{-19}$ C
k	= Coulomb constant[1]	$8.987\ 551\ 79 \times 10^8$ N·m²/C²
m_{oe}	= Electron rest mass[2]	$0.510\ 999\ 07$ MeV/c²
		$9.109\ 390 \times 10^{-31}$ kg
m_{on}	= Neutron rest mass[2]	$939.565\ 63$ MeV/c²
		$1.674\ 929 \times 10^{-27}$ kg
m_{op}	= Proton rest mass[2]	$938.272\ 31$ MeV/c²
		$1.672\ 623 \times 10^{-27}$ kg
$m_{o\mu}$	= Muon rest mass[2]	$105.658\ 389$ MeV/c²
$m_{o\tau}$	= Tau rest mass[2]	1777.00 MeV/c²
		$3.167\ 792 \times 10^{-27}$ kg
U	= Universal spiral field constant[*]	$137.035\ 988$
μ_B	= Bohr magneton[*]	$9.274\ 015 \times 10^{-24}$ C ·m²/s
μ_N	= Nuclear magneton[*]	$5.050\ 787 \times 10^{-27}$ C ·m²/s
μ_e/μ_B	= Electron magnetic moment[2]	$1.001\ 159\ 652\ 193$
μ_n/μ_N	= Neutron magnetic moment[2]	$-1.913\ 0428$
μ_n/μ_N	= Proton magnetic moment[2]	$2.792\ 847\ 39$
$\mu_\mu/\mu_{o\mu}$	= Muon magnetic moment[2]	$1.001\ 165\ 923$
Ω	= Gravitational electromass ratio[*]	$8.616\ 418 \times 10^{-11}$ C/kg.

Note: [*] derived values.

REFERENCES

1. Serway, R.A., *Physics For Scientists & Engineers with Modern Physics*, Third Edition, Saunders College Publishing, Philadelphia, 1992.
2. *Physical Review D: Particles and Fields*, Volume 54, Third Series, Part 1, Review of Particle Physics, The American Physical Society, 1 July, 1996.

APPENDIX B

A Unified Purpose For The Human Race
A Message From Helicola

Fundamental Questions - Since the beginning of conscious human life, we asked ourselves very profound questions: Why are we here? Are we a result of an accidental chemical reaction or there was a special purpose for our appearance in the Universe? And if there is a purpose then what is it?

Thanks to the efforts of many philosophers, scientists and theologians, we learned a great deal about the creation and evolution of human life and consciousness. However, the fundamental question about our purpose in the Universe still remains unanswered.

Think Cosmically - Fortunately, there is an alternative source of help. This help comes from Nature itself. Nature speaks to us in a language of pure logic and common sense. There is, however, a psychological barrier that we have to overcome. Nature exists on a cosmic scale, while we are mostly occupied with solving everyday problems on extremely small local scale, without seeing the big picture. Therefore, to understand Nature we must open our minds and begin to think cosmically.

Our Place in the Universe - Let us start by realizing that we are the participants in a great motion involving the entire Universe. Contrarily to what we were taught in schools, we are not flying apart as a result of the Big Bang. Neither are we moving with our planet Earth along annoyingly repetitive elliptical paths around the Sun. This would only be true if the Sun were to stand still, as Copernicus believed five centuries ago. Instead, we are moving along a multitude of ever enlarging celestial spiral paths wound around each other and forming an infinite multi-dimensional spiral shape called the helicola.

251

The Helicola – Together with our planet, we are moving along the first level of the helicola as the Earth rotates around its axis and moves around the Sun. Together with the Earth and the Sun, we are moving along the second level of the helicola as the Sun rotates around the center of our galaxy, the Milky Way, while our galaxy is moving in space. Together with the Earth, the Sun and the Milky Way, we are moving along the third level of the helicola as the Milky Way spirals around another entity that is still to be discovered, and so on and so on.

This cosmic vision helps us to realize that we are an integral part of the Universe. It also stimulates our inquiry about our role in the development of the Universe. As we will show below, this cosmic vision will eventually allow us to discover the role of the human race in the Universe by using the known facts about evolution of the Universe and by applying straightforward logic.

A Non-Adaptive Universe - During its long life, the Universe had evolved as a constantly enlarging sophisticated entity in accordance with the built-in algorithms that we call the laws of nature. These laws are universal, meaning that they are the same in all parts of the Universe. They are also immutable and unaffected by whether they are beneficial or detrimental to the existence of any particular part of the Universe. The detrimental effect of these laws, however, remained negligible until very recent times when the Universe became increasingly overcrowded, leading to more frequent collisions between not only stars but also between entire galaxies. Since celestial bodies do not have adaptive mechanisms, they have no means of recognizing the upcoming dangers and reacting rationally to avoid the disasters.

An Adaptive Living Universe - Fortunately, Nature was wise enough to recognize that a non-adaptive Universe was prone to self-destruction. Following the built-in law of self-preservation, She began well in advance a gradual process of conversion of a non-adaptive Universe into an adaptive Living Universe, that has a capability to sense, to think, and to respond rationally to upcoming events. As far as we know today, the only entity in the Universe that possesses this capability is the human race. This leads us to believe that Nature created the human race to play an indispensable role in creation of a Living Universe.

Preparing For Our Mission - Nature acted prudently and steadily. She created the utmost favorable conditions for the appearance of life and the human race on our planet. She provided us with sensors, brains and hands to perform the principal functions of adaptive systems. She provided us with the built-in algorithms for self-reproduction and self-preservation. She then gave us plenty of time to learn how to survive in this world by using natural resources. She strengthened our physical abilities by teaching us how to create and use machineries, and multiplied our thinking and communication skills by introducing computers and the Internet.

Earth became our classroom in which we were supposed to learn first how to create a Living Universe on the small scale of our planet, and then how to apply this knowledge to the entire Universe. It appears that Nature conducts this educational process on a timely basis, so we will certainly be ready to begin performing our mission when the non-adaptive Universe can no any longer survive without our help. It also appears that our participation in this process is not optional, but rather the law of nature that we all must carefully study and obey.

So far, we have not found anywhere in the Universe a civilization similar to ours. Is it possible that we are the first grain of a Living Universe? If so, then it gives us an even greater sense of responsibility for our mission.

Revising Our Thinking and Actions - Once Nature reveals to us the overwhelming importance of our role in the Universe, She wants us to begin revising our current thinking and actions that are frequently based on short-term, primitive and selfish goals. Our universal mission will serve as a benchmark against which we can measure the correctness of our everyday behavior. It will help young people to select the most appropriate friends, role models, and future professions. It will also allow them to avoid confusion when facing with contradictory advice from their parents, teachers, friends, spiritual leaders, and media.

Having our universal mission in mind will be useful for the adults as well, when handling everyday family and business affairs and solving extremely contradictory social, economic, and political problems. We will seek for the best qualities in each other that will help us to accomplish our common goal. We will consider every nation and every group of people as the partners in a common universal enterprise. We will all also begin to value differently our efforts in exploration of space.

Time For Our Mission Is Now - One may wonder how much time we have before starting our mission. Actually, the process of creating the Living Universe has already begun. It started from the moment when we became concerned about some global problems in our own planet that include social and economic inequalities between people, injustice, racial and religious prejudice, birth control, hunger, diseases, natural disasters, global warming, and destructive wars. Solving the problems in our own planet is the first step in creation of the Living Universe.

We are probably not making a good progress in solving all these problems. The absence of a strong guidance leads to various disagreements between us on how and at what pace we shall proceed. This may explain why Nature has decided that now was the right time to reveal the true purpose for the human race in the Universe This new knowledge will not only help us to unify our efforts, but also will make us confident that we will certainly succeed in this exciting cosmic endeavor because we follow one of the most important laws of nature that assures the survival and prosperity of the entire Universe.

LIST OF RELEVANT BOOKS CONSULTED

History & Biographies

Aaboe, A., *Episodes From The Early History of Mathematics*, Random House, 1972.

Agassi, J., *Faraday as a Natural Philosopher*, The University of Chicago Press, Chicago, ILL, 1971.

Aiton, E.J., *Leibniz-A Biography*, Adam Hilger Ltd, Bristol and Boston, 1985.

Alexandersson, O., *Living Water - Victor Schauberger and the Secrets of Natural Energy*, Gateway Books, Bath, UK, 1996.

Allen, R.E., *Greek Philosophy: Thales to Aristotle*, The Free Press, New York, 1966.

Alonzo, G.Z., *An Overview of the Mayan World*, Merida, Yucatan, Mexico, 1987.

Andrade, E.N. da C., *Rutherford and the Nature of the Atom*, Doubleday & Company, Inc., New York, 1964.

Beckmann, P., *A History of PI*, Dorset Press, New York, 1989.

Beiser, G., *The Story of Gravity - An Historical Approach to the Study of the Force That Holds the Universe Together*, E.P. Dutton & Co., Inc., New York , 1973.

Bethe, H.A., et al, *From a Life of Physics*, World Scientific Publishing Co., Singapore, 1989.

Beyerchen, A.D., *Scientists Under Hitler - Politics and the Physics Community in the Third Reich*, Yale University Press, New Haven, NJ, 1977.

Boorstin, D.J., *The Discoverers - A History of Man's Search to Know His World and Himself*, Random House, New York, 1983.

Born, M., *Thoughts and Remembrances of a Physicist*, "Nauka" Publishing, Moscow, 1977.

Boslough, J., *Stephen Hawking's Universe - An Introduction to the Most Remarkable Scientist of Our Time*, Quill/William Morrow, New York, 1985.

Bowers, B., *Michael Faraday and Electricity*, Priory Press Ltd., London, 1974.

Brinton, C., et al, *A History of Civilization 1300 to 1815*, Fifth Edition, Prentice Hall, Inc. Englewood Cliffs, New Jersey, 1987.

Caspar, M., *Kepler*, Abelard-Schuman, London and New York, 1959.

Ceram, C.W., *Gods, Graves, and Scholars - The Story of Archaeology*, Alfred A. Knopf, New York, 1964.

Cheney, M., *Tesla - Man Out of Time*, Prentice-Hall, Inc., Englewood Cliffs, N.J., 1981.

Clark, G., *The Man Who Tapped the Secrets of the Universe*, The University of Science and Philosophy, Swannanoa, Waynesboro, 2000.

Cohen, M.R. and Drabkin, I..E., *A Source Book in Greek Science*, McGraw-Hill Book Company, Inc., New York, 1948.

Coolidge, J.L., *The Mathematics of Great Amateurs*, Dover Publications, Inc., New York, 1963.

Crew, H., *The Wave Theory of Light - Memoirs by Huygens, Young anf Fresnel*, American Book Company, New York, 1900.

Cushing, J.T., *Philosophical Concepts in Physics - The Historical Relation Between Philosophy and Scientific Theories*, Cambridge University Press, Cambridge, UK, 1998.

De Broglie, L., *The Revolution in Physics*, The Noonday Press, New York, 1953.

De Broglie, L., *New Perspectives in Physics*, Basic Books, Inc., New York, 1962.

Dibner, B., *Oersted - And the Discovery of Electromagnetism*, Blaisdell Publishing Company, New York, 1962.

Diggins, J.E., *String, Straightedge and Shadow*, The Viking Press, New York, 1965.

Dijksterhuis, E.J., *Archimedes*, Princeton University Press, Princeton, N.J., 1987.

Domb, C., *Clerk Maxwell and Modern Science - Six Commemorative Lectures*, The Athlone Press, University of London, UK, 1963.

Drake, S., *Galileo at Work, His Scientific Biography*, The University of Chicago Press, Chicago. 1978.

Druce, J.G.F., *Brief Outline of the History of Science*, The Chemical News, Ltd., London, 1925.

Einstein, A., *Out of My Later Years*, A Citadel Press Book-Carol Publishing Group, New York, NY, 1991.

Einstein, A. and Infeld, L., *The Evolution of Physics - From Early*

Concepts to Relativity and Quanta, A Touchstone Book/Simon & Schuster, New York, 1966.

Eves, H., *An Introductory to the History of Mathematics*, Holt, Rinehart and Winston, New York, 1969.

Farrington, B., *Greek Science - Its Meaning for Us*, Penguin Books, Baltimore, MD, 1971.

Farrington, B., *Science and Politics in the Ancient World*, Barnes & Noble, Inc., New York, 1966.

Feder, T.H., *Great Treasures of Pompeii & Herculaneum*, Abbreville Press, Inc., Publishers, New York. 1985.

Frank, P., *Einstein - His Life and Times*, Da Capo Press, Inc., New York, 1947.

Gamow, G., *Thirty Years That Shook Physics - The Story of Quantum Theory*, Dover Publications, Inc., New York, 1966.

Gamow, G., *The Great Physicists From Galileo to Einstein*, Dover Publications, Inc., New York, 1961.

Gamow, G., *One, Two, Three...Infinity - Facts and Speculations of Science*, Dover Publications, Inc., New York, 1988.

Garbini, G., *The Ancient World*, McGraw-Hill Book Company, New York, 1993.

Geymonat, L., *Galileo Galilei: A Biography and Inquiry Into His Philosophy of Science*, McGraw-Hill Book Company, 1965.

Gillispie, C.C., *Dictionary of Scientific Biography*, Vol. IV, Charles Scribner's Sons, New York, 1971.

Grattan-Guinness, I., *Companion Encyclopedia of the History and Philosophy of the Mathematical Sciences, Vol.1 & Vol. 2*, Routledge, London and New York, 1994.

Gleick, J., *Genius - The Life and Science of Richard Feynman*, Pantheon Books, New York, 1992.

Gray, A., *Lord Kelvin - An Account of His Scientific Life and Work*, E.P. Dutton & Co., 1908.

Guillen, M., *Five Equations That Changed the World*, Hyperion, New York, NY, 1995.

Hadingham, E., *Early Man and Cosmos*, Walker and Company, New York, 1984.

Hall, A.R., *Isaac Newton - Adventurer in Thought*, Blackwell Publishers, Oxford, UK, 1994.

Heath, J.L., *The Works of Archimedes*, Dover Publications, Inc., New York, 1994.

Heilbron, J.L., *The Dilemmas of an Upright Man - Max Planck as Spokesman for German Science*, University of California Press, Berkeley, 1986.

Heindel, M., *Crucian Cosmo-Conception or Mystic Christianity*, W.B. Conkey Company, Hammond, Indiana, 1944.

Heisenberg, E., *Inner Exile - Recollections of a Life With Werner Heisenberg*, Birkhauser Boston, MA, 1980.

Hippel, F., *Citizen Scientist*, A Touchstone Book/Simon & Schuster, New York, 1991.

Hoffmann, B., *Albert Einstein - Creator and Rebel*, New American Library, New York, 1972.

Hurley, W.M., *Prehistoric Cordage*, Taraxacum Inc., Washington, 1979.

Ipsen, D.C., *Archimedes: Greatest Scientist of the Ancient World*, Enslow Publishers, Inc., Hillside, N.J., 1988.

Jones, B.Z., *The Golden Age of Science*, Simon and Schuster, New York, 1966.

Jonsson, I., *Emanuel Swedenborg*, Twayne Publishers Inc., New York, 1971.

Jung, C.G., *Man and His Symbols*, Bantam Doubleday Dell Publishing Group, Inc., New York, 1964.

Klauder, J.R., *Magic Without Magic: John Archibald Wheeler - A Collection of Essays in Honor of His Sixtieth Birthday*, W.H. Freeman and Company, San Francisco, CA, 1972.

Krupp, E.C., *In Search of Ancient Astronomies*, Doubleday & Company, Inc., Garden City, New York, 1977.

Krupp, E.C., *Echoes of the Ancient Skies - The Astronomy of Lost Civilizations*, Harper & Row, Publishers, New York, 1983.

Larsen, R., et al, *Emanuel Swedenborg - A Continuing Vision*, Swedenborg Foundation, Inc., New York, 1988.

Le Lionnais, F., *Great Currents of Mathematical Thought*, Dover Publications, Inc., New York, 1973.

Lenard, P., *Great Men of Science - A History of Scientific Progress*, G. Bell and Sons, Ltd., London, 1958.

Lerner R.G. and Trigg, G.L., *Encyclopedia of Physics*, Second Edition, VCH Publishers, Inc., New York, 1991.

Ludwig, C., *Michael Faraday - Father of Electronics*, Herald Press, Scottdale, PA, 1978.

Ludwig, E., *Michel Angelo*, Ernest Rowohlt Verlag, Berlin, 1930.

McClain, E.G., *The Pythagorean Plato - Prelude to the Song Itself,*

Nicolas Hays, Ltd., Stony Brook, New York, 1978.

Massie, R.K., *Peter The Great*, Alfred A. Knopf, New York, 1981.

Milton, R., *Alternative Science - Challenging the Myths of the Scientific Establishment*, Park Street Press, Rochester, Vermont, 1996.

Moore, W., *Schrodinger - Life and Thought*, Cambridge University Press, UK, 1992.

Munitz, M.K., *Theories of the Universe, From Babylonian Myth to Modern Science*, The Free Press, Glencoe, ILL, 1957.

Parry, A., *The Russian Scientist*, MacMillan Publishing Co., Inc., 1973.

Pogrebisskiy, J.B., *From Lagrange to Einstein*, (Russian), Yanus, Moscow, 1996.

Porter, R., *The Biographical Dictionary of Scientists*, Oxford University Press, New York, 1994.

Price, W.C., et al, *Wave Mechanics; The First Fifty Years - A Tribute to Professor Louis De Broglie*, John Wiley & Sons, New York-Toronto, 1973.

Rohrlich, F., *Classical Charged Particles - Foundations of Their Theory*, Addison-Wesley Publishing Company, Inc., Reading, MA, 1965.

Russell, B., *A History of Western Philosophy*, A Touchstone Book/Simon and Schuster, New York, 1945.

Sharlin, H.I., *Lord Kelvin - The Dynamic Victorian*, The Pennsylvania State University Press, PA, 1979.

Smith, D.E., *History of Mathematics*, Vol.1, Dover Publications, Inc., New York, 1964.

Smith, D.E., *History of Mathematics*, Vol.2, Dover Publications, Inc., New York, 1964.

Soderberg, H., *Swedenborg's 1714 Airplane - A Machine to Fly in the Air*, Swedenborg Foundation, New York, 1988.

Tesla, N., *Nicola Tesla:Lecture Before the New York Academy of Sciences - April 6, 1897*, Twenty First Century Books, Breckenridge, Colorado, 1994.

The History of Art: Architecture, Painting, Sculpture, Exeter Books, New York, 1985.

Time-Life Books, *The European Emergence - Time Frame AD 1500-1600*, The Time Inc. Book Company, Alexandria, VA .

Tricker, R.A.R., *The Contributions of Faraday and Maxwell to Electrical Science*, Pergamon Press Ltd., London, UK, 1966.

Van Nostrand's Scientific Encyclopedia, D. Van Nostrand Company, Inc., 1958.

Vassilatos, G., *Lost Science*, Adventures Unlimited Press, Kempton, ILL,

1999.

Vrooman, J.R., *Rene Descartes - A Biography*, G.P. Putnam's Sons, New York, 1970.

Van Der Waerden, B.L., *Science Awakening*, Science Editions, John Wiley & Sons, Inc., New York, 1963.

Westfall, R., *The Life of Isaac Newton*, Cambridge University Press, New York, NY, 1994.

Principal works

Boscovich, R.J., *A Theory of Natural Philosophy*, The M.I.T. Press, Cambridge, MA, 1966.

Dalton, J., et al, *Foundations of the Atomic Theory: Comprising Papers and Extracts*, Alembic Club, Edinburgh, UK, 1968.

French, A.P., *Newtonian Mechanics - The M.I.T. Introductory Physics Series*, W.W. Norton & Company, Inc., New York, 1971.

Lorentz, H.A., *Problems of Modern Physics - A Course of Lectures Delivered in the CA Institute of Technology*, Ginn and Company, Boston, 1927.

Newton, I., *The Principia*, Prometheus Books, Amherst, New York, 1995.

Niven, W.D., *The Scientific Papers of James Clerk Maxwell*, Dover Publications, Inc., New York, 1890.

Planck, M., *Eight Lectures on Theoretical Physics*, Dover Publications, Inc., New York, 1998.

Swedenborg, E., *The Principia*, Swedenborg Society, London, 1912.

Thomson, J.J., *Electricity and Matter*, Charles Scribner's Sons, New York, 1904.

Thomson, J.J., *Rays of Positive Electricity and Their Application to Chemical Analyses*, Longmans, Green and Co., London, UK, 1913.

Thomson, J.J., *A Treatise on Motion of Vortex Rings*, MacMillan and Co., London, 1883.

Wheeler, J.A., *Geometrodynamics, Topics of Modern Physics, Vol. 1*, Academic Press, Inc., New York, NY, 1962.

Mathematics

Adler, I., *Magic House of Numbers*, The John Day Company, New York, 1957.

Amir-Moez, A.R., *Ruler, Compass, and Fun*, Ginn and Company, 1967.

Anderson, H.B., *Analytic Geometry with Vectors*, McGutchan Publishing Corp., Berkeley, CA, 1984.

Arfken, G., *Mathematical Methods for Physicists*, Third Edition, Academic Press, Inc., New York, 1985.

Bak, T.A. and Lichtenberg, J., *Mathematics for Scientists*, W.A. Benjamin, Inc., 1966.

Batschelet, E., *Introduction to Mathematics for Life Scientists*, Springer-Verlag, New York, 1971.

Bronwell, A., *Advanced Mathematics in Physics and Engineering*, McGraw-Hill Book Company, Inc., 1953.

Carroll, R.L., *Simple Solutions to Impossible Problems*, the Carroll Research Institute, Columbia, S.C., 1986.

Cell, J.W., *Analytic Geometry*, John Wiley & Sons, Inc., New York, 1951.

Church, A.E. and Bartlett, G.M., *Elements of Descriptive Geometry*, American Book Company, New York, 1902.

Davis, P.J. and Hersh, R., *Descartes' Dream - The World According to Mathematics*, Harcourt Brace Jovanovich, Publishers, New York, 1986.

Dixon, R., *Mathographics*, Dover Publications, Inc., New York, 1987.

Domoryad, A.P., *Mathematical Games and Pastimes*, A Pergamon Press Book/The MacMillan Company, New York, 1964.

Downs, J.W., *Practical Conic Section*, Dale Seymour Publications, Palo Alto, CA, 1993.

Forder, H.G., *Geometry*, Hutchinson House, London, UK, 1950.

Gardner, M., *Mathematical Puzzles and Diversions*, Simon and Schuster, New York, 1959.

Gardner, M., *Mathematical Puzzles*, Thomas Y. Crowell Company, New York, 1977.

Gardner, M., *The Unexpected Hanging and Other Mathematical Diversions*, Simon and Schuster, New York, 1966.

Gardner, M., *Knotted Doughnuts and Other Mathematical Entertainments*, W.H. Freeman and Company, New York, 1986.

Gilmore, T., *Geocubic Cosmology*, T. Byron Publishing, Angels Camp, CA, 1993.

Gleick, J., *Chaos - Making a New Science*, Penquin Books, New York, 1988.

Grant, H.E., *Practical Descriptive Geometry*, McGraw-Hill Book Company, Inc., New York, 1952.

Gray, A., *Modern Differential Geometry of Curves and Surfaces*, CRC

Press, Boca Raton, 1993.

Henderson, L.D., *The Fourth Dimension and Non-Euclidean Geometry in Modern Art*, Princeton, 1983.

Hilbert, H. and Cohn-Vossen, S., *Geometry and the Imagination*, Chelsea Publishing Company, New York, 1952.

Hogben, L., *Mathematics in the Making*, Doubleday & Company, Garden City, New York, 1960.

Holt, M., *Mathematics in Art*, Van Nostrand Reinhold Company, New York, 1971.

Hooke, R., et al, *Math and Aftermath*, Walker and Company, New York, 1977.

Hunter, J.A.H. and Madachy, J.S., *Mathematical Diversions*, D. Van Nostrand Company, Inc., Princeton, New Jersey, 1968.

Kramer, E.E., *The Main Stream of Mathematics*, Fawcett Publications, Inc., Greenwich, Conn.,1972.

Lamit, L.G., *Descriptive Geometry*, Prentice-Hall, Inc., Englewood Cliffs, N.J., 1983.

Laugwitz, D., *Differential and Riemannian Geometry*, Academic Press, New York, 1965.

Lauwerier, H., *Fractals - Endlessly Repeated Geometrical Figures*, Princeton University Press, Princeton, New Jersey, 1991.

Lewis, H., *Geometry - A Contemporary Course*, McCormick-Mathers Publishing Company, Cincinnati, OH, 1973.

Lindgren, C.E., *Four - Dimensional Descriptive Geometry*, McGraw-Hill Book Company, New York, 1968.

Lockwood, E.H., *A Book of Curves*, Cambridge University Press, New York, 1961.

Lord, E.A. and Wilson, C.B., *The Mathematical Description of Shape and Form*, Ellis Horwood Ltd., Chichester, England, 1986.

Low, D. A., *Practical Geometry and Graphics*, Longmans, Green and Co., London, UK, 1995.

Marsden, J.E. and McCracken, M., *The HOPF Bifurcation and Its Application*, Springler-Verlag, New York, 1976 (Russian).

Middlemiss, et al, *Analytic Geometry*, McGraw-Hill Book Company, New York, 1986.

Ogilvy, C.S., *Excursion in Geometry*, Oxford University Press, 1969.

Pappas, T., *The Joy of Mathematics - Discovering Mathematics All Around You*, Wide World Publishing/Tetra, San Carlos, CA, 1989.

Ravielli, A., *An Adventure In Geometry*, The Viking Press, New York,

1957.

Resnikoff, H. L., *Mathematics in Civilization,* Holt, Rinehart and Winston, Inc., New York, 1974.

Rowe, C.E. and McFarland, J.D., *Engineering Descriptive Geometry,* D. Van Nostrand Company, Inc., New York, 1939.

Rucker, R., *Mind Tools - The Five Levels of Mathematical Reality,* Houghton Miffin Company, Boston, 1987.

Schumann, C.H., *Descriptive Geometry,* D. Van Nostrand Company, Inc., New York 1927.

Seggern, D., *CRC Standard Curves and Surfaces,* CRC Press, Boca Raton, Florida, 1993.

Seife, C., *Zero - The Biography of a Dangerous Idea,* Viking Penquin, New York, 2000.

Slagle, W.C.H., *Descriptive Geometry,* McGraw-Hill Book Co., New York, 1912.

Smith, K.J., *The Nature of Mathematics,* Brooks/Cole Publishing Company, Monterey, CA, 1984.

Smullyan, R.M., *Forever Undecided - A Puzzle Guide to Gödel,* Alfred A. Knopf, New York, 1987.

Sprott, J.C., *Strange Attractions - Creating Patterns in Chaos,* M&T Books, New York, 1993.

Steen, L.A., *Mathematics Today - Twelve Informal Essays,* Vintage Books/Random House, New York, 1980.

Steinhaus, H., *Mathematical Snapshots,* Oxford University Press, New York, 1960.

Terry, L., *The Mathmen,* McGraw-Hill Book Company, New York, 1964.

Wolfram, S., *A New Kind of Science,* Wolfram Media, Inc. Champaign, IL, 2002.

Zwikker, C., *The Advanced Geometry of Plane Curves and Their Applications,* Dover Publications, Inc., New York, 1994.

Spirals & Srings in Theories of Physics

Aiton, E.J., *The Vortex Theory of Planetary Motions,* American Elsevier, Inc., New York, 1972.

Carter, J., *The Other Theory of Physics-A Non-Field Unified Theory of Matter and Motion,* Absolute Motion Press, 2000.

Davies, P.C.W. and Brown, J., *Superstrings - A Theory of Everything?,*

Cambridge University Press, Cambridge, UK, 1988.

Ginzburg, V.B., *Spiral Grain of the Universe-In Search of the Archimedes File*, University Editions, Inc., Huntington, WV, 1996.

Ginzburg, V.B., *Unified Spiral Field and Matter - A Story of a Great Discovery*, Helicola Press, Pittsburgh, PA, 1999.

Greene, B., *The Elegant Universe - Superstrings, Hidden Dimensions, and the Quest for the Ultimate Theory*, W.W. Norton & Company, New York, 1999.

Kaku, M., *Introduction to Superstrings*, Springer-Verlag, New York, 1988.

Peat, F.D., *Superstrings and the Search for The Theory of Everything*, Contemporary Books, Lincolnwood (Chicago), IL, 1988.

Winter, D., at el, *Alphabet of the Heart-Sacred Geometry: The Genesis in Principle of Language and Feeling*, Eden, New York, 1994.

Nuclear & Atomic physics

Adair, R.K., *The Great Design-Particles, Fields and Creation*, Oxford University Press, Oxford, UK, 1987.

Albert, D.Z., *Quantum Mechanics and Experience*, Harvard University Press, Cambridge, MA, 1992.

Asimov, I., *The Neutrino - Ghost Particle of the Atom*, Doubleday & Company, Inc., Garden City, NY, 1966.

Baeyer, H.C., *Rainbows, Snowflakes, and Quarks-Physics and the World Around Us*, McGraw-Hill Book Company, New York, 1984.

Baggott, J., *The Meaning of Quantum Theory*, Oxford University Press, Oxford, UK, 1992.

Bennet, A., et al, *Electrons on the Move*, Walker and Company, New York, 1964.

Blackwood, O.H., et al, *An Outline of Atomic Physics*, John Wiley & Sons, Inc., New York, 1955.

Blokhintsev, D.I., *Quantum Mechanics*, D.Reidel Publishing Company, Dordrecht, Holland, 1964.

Born, M., *Atomic Physics*, Dover Publications, Inc., Mineola, NY, 1989.

Brennan, R.P., *Heisenberg Probably Slept Here*, John Wiley & Sons, Inc., 1997.

Butler, S.T. and Messel, H., *Space and the Atom - A Course of Selected Lectures in Physics and Astronomy*, University of Sydney, 1961.

Carrigan, R.A. and Trower, W.P., *Particles and Forces at the Heart of* *the Matter, Readings from Scientific American*, W.H. Freeman and Company, New York, 1990.

Carrigan, R.A. and Trower, W.P., *Particle Physics in the Cosmos, Readings from Scientific American*, W.H. Freeman and Company, New York, 1989.

Cline, B.L., *The Questioners: Physicists and the Quantum Theory*, Thomas Y. Crowell Company, New York, 1965.

Close, F., et al, C., *The Particle Explosion*, Oxford University Press, New York, 1994.

Day, W., *A New Physics*, Foundation for New Directions, Cambridge, MA, 2000.

Dodd, J.E., *The Ideas of Particle Physics - An Introduction for Scientists*, Cambridge University Press, Cambridge, UK, 1984.

Dunning, F.B. and Hulet, R.G., *Atomic, Molecular, and Optical Physics:Charged Particles*, Academic Press, New York, 1995.

Enge, H.A., *Introduction to Nuclear Physics*, Addison-Wesley Publishing Company, Inc., Reading, MA, 1966.

Epstein, L.C., *Thinking Physics Is Gedanken Physics*, Insight Press, San Francisco, CA, 1983.

Feynman, R.P., *Six Easy Pieces and Six Not-So-Easy Pieces*, Perseus Publishing, Cambridge, MA, 1997.

Frauenfelder, H. and Henley, E.M., *Subatomic Physics*, Prentice-Hall, Englewood Cliffs, NJ, 1974.

Fritzsch, H., *Quarks - The Stuff of Matter*, Basic Books, Inc., New York, 1983.

Gautreau, R. and Savin, W., *Modern Physics*, McGraw-Hill, New York, 1978.

Graser, F.X., *The A-B-C of Electrons, Atoms, and Molecules*, Greenwich Book Publishers, New York, 1957.

Gribbin, J., *Schrodinger's Kittens and the Search for Reality*, Little, Brown and Company, Boston, New York, Toronto, London, 1995.

Gribbin, J., *Q Is For Quantum*, An Encyclopedia of Particle Physics, A Touchstone Book, Simon & Schuster, New York, 2000.

Griffiths, D., *Introduction to Elementary Particles*, Harper & Row, Publishers, New York, 1987.

Herzberg, G., *Atomic Spectra and Atomic Structure*, Dover Publications, Inc., New York, 1944.

Hill, R.D., *Tracking Down Particles*, W.A. Benjamin, Inc., New York, 1964.

Honig, W.M., *The Quantum and Beyond*, Philosophical Library, New York, 1986.

Honig, W.M., et al, *Quantum Uncertainties - Recent and Future Experiments and Interpretations*, Plenum Press, New York and London, 1987.

Hooft, G., *In Search of the Ultimate Building Blocks*, Cambridge University Press, UK, 1997.

Hughes, I.S., *Elementary Particles*, Cambridge University Press, Cambridge, UK, 1972.

Hughston, L.P., *Lecture Notes in Physics - Twistors and Particles*, Springer-Verlag, New York, 1979.

Jackson, J.D., *The Physics of Elementary Particles*, Princeton University Press, Princeton, NJ, 1958.

Kane, G., *The Particle Garden - Our Universe as Understood by Particle Physicists*, Addison-Wesley Publishing Company, Reading, MA, 1995.

Kuhn, T.S., *Black-Body Theory and the Quantum Discontinuity, 1894-1912*, The University of Chicago Press, Chicago, 1987.

Kenyon, I.R., *Elementary Particle Physics*, Routledge & Kegan Paul, London, UK, 1987.

Kostro, L., et al, *Problems in Quantum Physics; Gdansk '87 - Recent and Future Experiments and Interpretations*, World Scientific, New Jersey, 1987.

Lederman, L., *The God Particle*, Bantam Doubleday Dell Publishing Group, Inc., New York, 1993.

Liebschner, H., *Technical Fundamentals - Isotopes in Research and Production*, Edition Leipzig, GDR, 1969.

Livingston, M.S., *Particle Physics - The High-Energy Frontier*, McGraw-Hill Book Company, New York, 1968.

Lockyer, T.N., *Vector Particle Physics*, TNL Press, Los Altos, CA, 1991.

Massey, H., *The New Age in Physics*, Harper & Brothers, Publishers, New York, 1961.

Mott, N.F., *Elements of Wave Mechanics*, Cambridge University Press, London, UK, 1952.

Physical Review D: Particles and Fields, Vol. 54, The American Physical Society, 1996.

Polkinghorne, J.C., *The Quantum World*, Princeton University Press, Princeton, NJ, 1984.

Riordan, M., *The Hunting of the Quark-A True Story of Modern Physics*, Simon and Schuster/Touchstone, New York, 1987.

Segre, E., *Nuclei and Particles - An Introduction to Nuclear and Subnuclear Physics*, W.A. Benjamin, Inc., New York, 1965.

Semat, H., *Introduction to Atomic and Nuclear Physics*, Rinehart & Company, Inc., New York, 1958.

Semat, H. and Katz, R., *Physics*, Rinehart & Company, Inc., New York, 1987.

Serway, R.A., *Physics - For Scientist & Engineers with Modern Physics*, Saunders College Publishing, Philadelphia,1992.

Shankland, R.S., *Atomic and Nuclear Physics*, The Macmillan Company, New York, 1960.

Sproull, R.L., *Modern Physics - A Textbook for Engineers*, John Wiley & Sons, Inc., New York, 1956.

Stearns, R.L., *Basic Concepts of Nuclear Physics*, Reinhold Book Corporation, New York, 1968.

Stewart, A.T., *Perpetual Motion - Electrons and Atoms in Crystals*, Anchor Books/Doubleday & Company, Inc., New York, 1965.

Stranathan, J.D., *The "Particles" of Modern Physics*, The Blakiston Company, Philadelphia, 1942.

Stwertka, A., *Recent Revolutions in Physics - The Subatomic World*, Franklin Watts, New York, 1985.

Swartz, C.E., *The Fundamental Particles*, Addison-Wesley Publishing Company, Inc., Reading, MA, 1965.

Tomkeiffe, S.I., *A New Periodic Table of the Elements Based on the Structure of the Atom*, Chapman & Hall, Ltd., 1954.

Venable, W.M., *The Interpretation of Spectra*, Reinhold Publishing Corporation, New York, 1948.

Vlasov, A.D., *Theory of Linear Accelerators*, Atomizdat, Moskva, 1965.

Weinberg, S., *Dreams of a Final Theory*, Pantheon Books, New York, 1992.

White, H.E., *Introduction to Atomic Spectra*, McGraw-Hill Book Company, Inc., New York, 1934.

Williams, J.E., *Modern Physics*, Holt, Rinehart and Winston, Publishers, New York, 1984.

Wilson, R.R. and Littauer, R., *Accelerators - Machines of Nuclear Physics*, Anchor Books/Doubleday & Company, Inc., New York, 1960.

Wolff, M., *Exploring The Physics of Unknown Universe*, Nova Science Publishers, Inc.Commack, NY, 1990.

Astronomy, Cosmology, Space & Time

Alfven, H., *Worlds - Antiworlds: Antimatter in Cosmology*, W.H. Freeman and Company, San Francisco, 1966.

Alfven, H. and Arrehenius, G., *Evolution of the Solar System*, (Russian), "Mir" Publishers, Moscow, 1979.

Arp, H., *Seeing Red - Redshifts, Cosmology and Academic Science*, Apeiron, Montreal, Canada, 1998.

Barrow, J.D. and Silk, J., *The Left Hand of Creation*, Oxford University Press, New York, 1983.

Bloyd, J.G., *Broken Arrow of Time-Rethinking the Revolution in Modern Physics*, Writers Club Press, San Jose, CA, 2001.

Born, M., *The Restless Universe*, Dover Publications, Inc., New York, 1951.

Boslough, J., *Masters of Time - Cosmology at the End of Innocence*, Addison-Wesley Publishing Company, Reading, MA, 1992.

Bova, B., *Closeup: New Worlds*, St. Martin's Press, New York, 1977.

Butler, S.T. and Messel, H., *Space and the Atom - A Course of Selected Lectures in Physics and Astronomy*, University of Sydney, 1961.

Byron, T.G., *The Geocubic Matrix in the Universe and the Cosmos of Energy/Matter Caught in its Time-Flow*, T Byron G Publishing, Angels Camp, CA, 1992.

Caes, C.J., *Cosmology-The Search for the Order of the Universe*, Tab Books Inc., Blue Ridge Summit, PA, 1986.

Carroll, R.L., *The Energy of Physical Creation*, the Carroll Research Institute, Columbia, S.C., 1985.

Davies, P., *About Time - Einstein's Unfinished Revolution*, Simon & Schuster, New York, 1995.

Dixon, R.T., *Dynamic Astronomy*, Sixth Edition, Prentice-Hall, Inc., Englewood Cliffs, NJ, 1991.

Driscoll, R.B., *United Theory of Ether, Field and Matter*, Published by Author, Portland, OR, 1964.

Driscoll, R.B., *United Theory of Ether, Field and Matter (supplement)*, Published by Author, Oakland, CA, 1965.

Duncan, J.C., *Astronomy - A Textbook*, Harper & Brothers Publishers, New York, 1926.

Eckhart, L., *Four-Dimensional Space*, Indiana University Press, Bloomington, 1968.

Flood, R. and Lockwood, M., *The Nature of Time*, Basil Blackwell, Inc.,

Cambridge, MA., 1990.

Fraser, G., et al, *The Search for Infinity*, Facts on File, Inc., New York, 1995.

Fritzsch, H., *The Creation of Matter - The Universe From Beginning to End*, Basic Books, Inc., New York, 1984.

Genz, H., *Nothingness - The Science of Empty Space*, Perseus Books, Reading, MA, 1999.

Harrison, E.R., *Cosmology - The Science of the Universe*, Cambridge University Press, Cambridge, UK, 1981.

Harrison, E.R., *Masks of the Universe*, Macmillan Publishing Company, New York, 1985.

Hawking, S., *Black Holes and Baby Universes and Other Essays*, Bantam Books, New York, 1993.

Hawking, S., *A Brief History of Time - From the Big Bang to Black Holes*, Bantam Books, New York, 1990.

Hawking S. and Penrose, R., *The Nature of Space and Time*, Princeton University Press, Princeton, NJ, 2000.

Jaki, S.L., *Planets and Planetarians: A History of Theories of the Origin of Planetary Systems*, John Wiley & Sons, New York, 1979.

Kaku, M., *Hyperspace*, Anchor Books/Doubleday, New York, 1994.

Kaku, M., *Beyond Einstein - The Cosmic Quest for Theory of the Universe*, Anchor Books/Doubleday, New York, 1995.

Krauss, L., *Quintessence-The Mystery of Missing Mass in the Universe*, Basic Books, New York, NY, 2000.

Layzer, D., *Constructing the Universe*, Scientific American Books, Inc., New York, 1984.

Lerner, E.J., *The Big Bang Never Happened*, Times Books, Random House, 1992.

Lindley, D., *The End of Physics - The Myth of a United Theory*, HarperCollins Publishers, Inc., 1993.

Manning, H.P., *The Fourth Dimension Simply Explained*, Dover Publications, Inc., New York, 1960.

Novikov, I.D., *The River of Time*, Cambridge University Press, Cambridge, UK, 1998.

Pagels, H.R., *The Cosmic Code - Quantum Physics as the Language of Nature*, Bantam Books, New York, 1982.

Parker, B., *Creation: the Story of the Origin and Evolution of the Universe*, Plenum Press, New York, 1988.

Peebles, P.J.E., *Principles of Physical Cosmology*, Princeton University

Press, Princeton, N.J., 1993.

Price, H., *Time's Arrow and Archimedes' Point*, Oxford University Press, New York, 1996.

Ridley, B.K., *Time, Space and Things*, Cambridge University Press, Cambridge, UK, 1994.

Riordan, M. and Schramm, D.N., *The Shadows of Creation - Dark Matter and the Structure of the Universe*, W.H. Freeman and Company, New York, 1991.

Rucker, R., *The Fourth Dimension*, Houghton Mifflin Company, Boston, 1984.

Russell, P., *The White Hole in Time*, Harper San Francisco, 1992.

Sagan, C., *Cosmos*, Ballantine Books, New York, 1980.

Schneider, M.S., *A Beginner's Guide to Constructing The Universe*, HarperPerennial, New York, 1995.

Staguhn, G., *God's Laughter - Man and His Cosmos*, HarperCollins, New York, 1992.

Sternglass, E.J., *Before the Big Bang - The Origins of the Universe*, Four Walls Eight Windows, New York, NY, 1997.

Thiel, R., *And There Was Light - The Discovery of the Universe*, Alfred A. Knopf, New York, 1957.

Skilling, W.T. and Richardson, R.S., *A Brief Text in Astronomy*, A Holt-Dryden Book/Henry Holt and Company, 1959.

Thorne, K.S., *Black Holes & Time Warps-Einstein's Outrageous Legacy*, W.W. Norton & Company, New York, 1994.

Van Flandern, T., *Dark Matter, Missing Planets & New Comets-Paradoxes Resolved, Origins Illuminated*, North Atlantic Books, Berkeley, CA, 1993.

Wheeler, J.A., *Geons, Black Holes & Quantum Foam - A Life in Physics*, W.W. Norton and Company, New York, NY, 1998.

Wolf, F.A., *Parallel Universes - The Search for Other Worlds*, A Touchstone Book, New York, 1990.

Zombeck, M.V., *Handbook of Space Astronomy and Astrophysics*, Cambridge University Press, Cambridge, UK, 1990.

Gravity, Inertia & Mass

Gamow, G., *Gravity*, Anchor Books/Doubleday & Company, Inc., 1962.

Ghosh, A., *Origin of Inertia*, Apeiron, Montreal, Canada, 2000.

Jammer, M., *Concepts of Mass in Classical and Modern Physics*, Harvard University Press, Cambridge, MA, 1961.

Jammer, M., *Concepts of Mass in Classical and Modern Physics*, Dover Publications, Inc., Mineola, New York, 1997.

Jammer, M., *Concepts of Force*, Dover Publications, Inc., Mineola, New York, 1999.

Klyushin, J.G., *On the Maxwell Approach to Gravity*, St. Petersburg, Russia, 1995.

Misner, C.W., et al, *Gravitation*, W.H. Freeman and Company, San Francisco, CA, 1973.

Nardo, D., *Gravity, the Universal Force*, Lucent Books, San Diego, CA, 1995.

Rudnicki, K., *Gravitation, Electromagnetism and Cosmology - Toward a New Synthesis*, Apeiron, Montreal, Canada, 2001.

Salem, K.G., *The New Gravity - A New Force - A New Mass - A New Acceleration - Unifying Gravity with Light*, Salem Books, Johnstown, PA, 1994.

Valens, E.G., *The Attractive Universe: Gravity and Shape of Space*, Motion, Magnet, 1969.

Relativity

Condon, E.U. and Odishaw, H., *Handbook of Physics*, McGraw-Hill Book Company, New York, 1967.

Einstein, A., *Relativity - The Special and the Clear Theory*, Crown Publishers, Inc., New York, 1961.

Epstein, L.C., *Relativity Visualized*, Insight Press, San Francisco, CA, 1992.

French, A.P., *Special Relativity - The M.I.T. Introductory Physics Series*, W.W. Norton and Company, Inc., New York, 1968.

Gardner, M., *Relativity for the Million*, Pocket Books, Inc., New York, 1967.

Hoffmann, B., *Relativity and Its Roots*, (Russian), Znaniye, Moscow, 1987.

McGervey, J.D., *Untroduction to Modern Physics*, Academic Press, New York, 1983.

Miller A.I., *Albert Einstein's Special Theory of Relativity*, Springer-Verlag New York, Inc., 1998.

Ney, E.P., *Electromagnetism and Relativity*, Harper & Row, Publishers, New York, 1962.

Pauli, W., *Theory of Relativity*, Pergamon Press, London, UK, 1958.

Rindler, W., *Essential Relativity - Special, General, and Cosmology*, Van Nostrand Reinhold Company, New York, 1969.

Rosser, W.G.V., *An Introduction to The Theory of Relativity*, Butterworths, London, 1964.

Rosser, W.G.V., *Classical Electromagnetism Via Relativity - An Alternative Approach to Maxwell's Equations*, Plenum Press, New York, 1968.

Rosser, W.G.V. and McCulloch, R.K., *Relativity and High Energy Physics*, Wykeham Publications, Ltd., London, 1969.

Sciama, D.W., *The Physical Foundations of General Relativity*, Doubleday & Company, Inc., 1969.

Sears, F.W., et al, *University Physics,* Seventh Edition, Addison-Wesley Publishing Company, Reading, MA, 1987.

Skinner, R., *Relativity*, Blaisdell Publishing Company, Waltham, Massachusetts, 1970.

Swisher, C., *Great Mysteries - Relativity: Opposing Viewpoints*, Greenhaven Press, Inc., San Diego, CA, 1990.

Nature

Ayensu, E., and Whitfield, P., *The Rhythms of Life*, Crown Publishers, Inc., New York, 1982.

Coats, C., *Living Energies*, Gateway Books, Bath, UK, 1996.

Cook, T.A., *The Curves of Life*, Dover Publications, Inc., New York, 1979.

Davies, P. and Gribbin, J., *The Matter Myth - Dramatic Discoveries That Challenge Our Understanding of Physical Reality*, Touchstone Book/Simon & Schuster, New York, 1992.

DeBusk, A.G., *Molecular Genetics*, McMillan Company, New York, 1969.

Engel, L., *The New Genetics*, Doubleday & Company, Inc., Garden City, New York, 1967.

Fast, J., *Blueprint for Life - The Story of Modern Genetics*, St. Martin's Press, New York, 1964.

Gardner, M., *The Ambidextrous Universe*, Basic Books, Inc., New York, 1964.

Gottlieb, M., et al, *Seven States of Matter - A Westinghouse Search Book*,

Walker and Company, New York, 1966.

Hamilton, W.R., et al, *A Guide to Minerals, Rocks and Fossils*, Crescent Books, New York, 1974.

Hargittai, I. and Pickover, C.A., *Spiral Symmetry*, World Scientific Publishing Co., Singapore, 1992.

Harrison, L.P., *Meteorology*, National Aeronautics Council, Inc., New York, 1942.

Lindbergh, A.M., *Gift From the Sea*, Pantheon Books, Division of Random House, Inc., New York, NY, 1955.

Lugt, H.J., *Vortex Flow in Nature and Technology*, Krieger Publishing Company, Malabar, Florida, 1995.

McCoy, C.J., *Identification Guide to Pennsylvania Snakes*, Educational Bulletin Number 1, Trustees of Carnegie Institute, 1980.

Morrison, P., *Powers of Ten: A Book About The Relative Size of Things in the Universe and the Effect of Adding Another Zero*, Scientific American Books, Inc., New York, 1982.

Murchie, G., *The Seven Mysteries of Life - An Exploration in Science and Philosophy*, Houghton Mifflin Company, Boston, 1978.

Murchie, G., *Music of the Spheres: The Material Universe - From Atom to Quasar, Simply Explained*, Dover Publications, Inc., New York, 1961.

Stent, G.S., *Molecular Genetics-An Introductory Narrative*, W.H. Freeman and Company, San Francisco, CA, 1974.

Scientific American, *The Enigma of Weather*, Scientific American, New York, NY, 1994.

Taylor, B.B., *Design Lessons From Nature*, Watson-Guptill Publications, New York, 1974.

The Illustrated Library of Nature, *Nature Hobbies*, Vol.10, H.S. Stuttman Co., Inc., 1971.

Uman, M.A., *Lightning*, Dover Publications, Inc., New York, NY, 1969.

Watson, J.D., *The Double Helix*, W.W. Norton & Company, New York, 1980.

Young, L.B., *The Mystery of Matter*, Oxford University Press, New York, 1965.

Zimmerman, B.E. and Zimmerman, D.J., *Why Nothing Can Travel faster Than Light...and Other Explorations in Nature's Curiosity Shop*, NTC/Contemporary Books, Lincolnwood (Chicago), 1993.

World Atlas, *The Phenomena*, The Curtis-Doubleday World Atlas, 1962.

Technology

Begich, N. and Manning, J., *Angels Don't Play This Haarp - Advances in Tesla Technology*, Earthpulse Press, Anchorage, Alaska, 1995.

Brown, R.J., *333 Science Tricks & Experiments*, Tab Books INC., Blue Ridge Summit, PA, 1984.

Bunch, B., *Handbook of Current Science & Technology*, Gale, New York, USA, 1996.

Burke, H.E., *Handbook of Magnetic Phenomena*, Van Nostrand Reinhold Company, New York, 1986.

Campbell, R.W., *Tops and Gyroscopes*, Thomas Y. Crowell Company, New York, 1961.

Cole, E.B., *The Theory of Vibrations for Engineers*, Crosby Lockwood & Son, Ltd., UK, 1950.

Crandall, B.C., *Nanotechnology - Molecular Speculations on Global Abundance*, The MIT Press, Cambridge, MA, 1996.

Davenport, E.G., *Your Handspinning*, Craft & Hobby Book Service, Big Sur, CA, 1964.

Ford, R.A., *Tesla Coil Secrets - Construction Notes and Novel Uses*, Lindsay Publications, Inc., Bradley, ILL, 1985.

Ginzburg, V.B., "Continuous Spiral Motion System for Roling Mills," U.S. Patent No. 5,970,771, October 26, 1999.

Graumont, R. and Hensel, J., *Splicing Wire and Fiber Rope*, Cornell Maritime Press, Cambridge, MA, 1955.

Green, R.M., *Commentary on the Effect of Electricity on Muscular Motion*, Elizabeth Licht, Publisher, Cambridge, MA, 1953.

Haag, J., *Oscillatory Motions*, Wadsworth Publishing Company, Inc., Belmont, CA, 1967.

Hambidge, J., *The Elements of Dynamic Symmetry*, Dover Publications, Inc., New York, 1967.

Kemp, B. and Azteca, I., *ABC's of Magnetism*, Howard W. Sams & Co., Inc., New York, 1962.

Kennedy, T., *Fun With Electricity*, Gernsback Library, Inc., New York, 1961.

Knight, D.C., *The Science Book of Meteorology*, Franklin Watts, Inc., New York, 1964.

Louisell, W.H., *Coupled Mode and Parametric Electronics*, John Wiley & Sons, Inc., 1960.

Lugt, H.J., *Vortex Flow in Nature and Technology*, Krieger Publishing

Company, Malabar, Florida, 1995.

Perry, R.H. and Chilton, C.H., *Chemical Engineers' Handbook*, McGraw-Hill Book Company, New York , 1973.

Ponamorev, C.D. and Andreeva, L.E., *Calculation of The Elastic Elements of Machines and Intrumentation*, Machinostroenie, Moscow, 1980 (Russian).

Scientific American, *The Origins of Technology*, Scientific American, 1997.

Shigley, J.E. and Mischke, C.R., *Standard Handbook of Machine Design*, McGraw-Hill Book Company, New York, 1986.

Siegel, M.J., et al, *Mechanical Design of Machines*, International Textbook Company, Scranton, PA, 1965.

Tennenbaum, J., *The Coming Breakthroughs in Biophysics and Mathematics*, Proceedings of the Krafft A. Ehricke Memorial Conference "Colonize Space! Open the Age of Reason, New Benjamin Franklin House, New York, 1985.

Timoshenko, S., *Strength of Materials*, Third Edition, D. Van Nostrand Company, Inc., Princeton, NJ, Toronto, London, 1956.

Tufte, E.R., *Envisioning Information*, Graphics Press, Cheshire, Connecticut, 1992.

Watkins Cyclopedia of the Steel Industry, Thirteenth Edition, Steel Publications, Inc., Pittsburgh, PA, 1971.

Wilson, F.W., *Tool Engineers Handbook*, McGraw-Hill Book Company, Inc., New York, 1949.

Plasma & Fluids

Bradshaw, P., *Experimental Fluid Mechanics*, Pergamon Press, Oxford, UK, 1964.

Dodge, R.A. and Thompson, M. J., *Fluid Mechanics*, McGraw-Hill Book Company, Inc., New York, 1937.

Ericsson, J., *Propelling Steam Vessels*, U.S. Patent # 588, February 1, 1838.

Eskinazi, S., *Vector Mechanics of Fluids and Magnetofluids*, Academic Press, New York, 1967.

Goldstein, S., *Modern Developments in Fluid Dynamics*, Oxford at the Clarendon Press, UK, 1960.

Loitsyanskii, L.G., *Mechnics of Liquids and Gases*, Pergamon Press,

Oxford, UK, 1966.

Massey, B.S., *Mechanics of Fluids*, D. Van Nostrand Company Ltd., London, UK, 1969.

Physics Survey Committee, *Physics in Perspective: The Nature of Physics and the Subfields of Physics,* Student Edition, National Academy of Sciences, Washington, DC, 1973.

Physics Survey Committee, *Plasma and Fluids - Physics Through the 1990's*, National Academy Press, Washington, DC, 1986.

Priyani, V. B., *The Fundamental Principles of Hydraulics*, Vol.1, Charotar Book Stall, Anand, India, 1957.

Rouse, H., *Fluid Mechanics for Hydraulic Engineers*, Dover Publications, Inc., New York, 1961.

Science & Consciousness

Aquarian, A., *The Little Scroll*, P.E.A.C.E Publications ehf, Iceland, 2001.

Ash, D. and Hewitt, P., *The Vortex - Key to Future Science*, Gateway Books, Bath, England, 1991.

Brophy, T., *The Mechanism Demands A Mysticism*, Medicine Bear Publishing, Blue Hill, ME, 1999.

Capra, F., *The TAO of Physics*, Shambhala, Boston, MA, 2000.

Chaisson, E., *The Life Era - Cosmic Selection and Conscious Evolution*, The Atlantic Monthly Press, New York, 1987.

Friedman, N., *Bridging Science and Spirit - Common Elements in David Bohm's Rhysics, The Perennial Philosophy and Seth*, Living Lake Books, St. Louis, MO, 1994.

Goswami, A., *The Self-Aware Universe - How Consciousness Creates the Material World*, Penquin Putnam Inc., 1993.

Kafatos, M. and Nadeau, R., *The Conscious Universe - Part and Whole in Modern Physical Theory*, Springer-Verlag New York, Inc., New York, 1990.

Penrose, R., *The Emperor's New Mind-Concerning Computers, Minds, and the Law of Physics*, Oxford University Press, New York, Oxford, 1989.

Penrose, R., *Shadows of the Mind - A Search for the Missing Science of Conciousness*, Oxford University Press, New York, 1994.

Rubik, B., *The Interrelation Between Mind and Matter*, Proceedings of a Conference Hosted by the Center for Frontier Sciences at Temple University, Philadelphia, PA, 1992.

Russell, W., *The Universal One*, The University of Science and Philosophy, Swannanoa, Waynesboro, VA, 1974.

Russell, W., *A New Concept of the Universe*, The University of Science and Philosophy, Swannanoa, Waynesboro, VA, 1989.

Russell, W., *The Secret of Light*, The University of Science and Philosophy, Swannanoa, Waynesboro, VA, 1994.

Sanders, P.A. Jr., *Scientific Vortex Information*, Free Soul Publishing, Sedona, AZ, 1992.

Weitzman, K., *Age of Spirituality: Late Antique and Early Christian Art, Third to Seventh Century*, Princeton University Press, MA, 1992.

Wolf, F.A., *The Eagle's Quest - A Physicist Search for Truth in the Heart of the Shamanic World*, A Touchstone Book/Simon & Schuster, New York, 1991.

Wyckoff, J., *Franz Anton Mesmer - Between God and Devil*, Prentice-Hall, Inc., Englewood Cliffs, NJ, 1975.

New Energy

Begich, N., *Towards a New Alchemy-The Millenium Science*, Earthpulse Press, Anchorage, Alaska, 1996.

Cathie, B.L., *The Bridge to Infinity - Harmonic 371244*, America West Publishers, Bozeman, MT, 1989.

Cathie, B.L., *The Energy Grid - Harmonic 695, The Pulse of the Universe*, America West Publishers, Bozeman, MT, 1990.

Cathie, B.L., *The Harmonic Conquest of Space*, Nexus Magazine, Mapleton, Queensland, Australia, 1995.

Childress, H., *The Free-Energy Device Handbook - A Compilation of Patents & Reports*, Adventures Unlimited Press, USA, 1994.

Collins, H. and Pinch, T., *The Golem - What Everyone Should Know About Science*, Cambridge University Press, Melbourne, Australia, 1996.

King, M. B., *Tapping The Zero-Point Energy*, Paraclete Publishing, Provo, UT, 1989.

King, M. B., *Quest For Zero-Point Energy - Engineering Principles For "Free Energy,"* Paraclete Publishing, Provo, UT, 2001.

Cook, N., *The Hunt For Zero Point - Inside The Classified World of Antigravity Technology*, Broadway Books, New York, 2001.

Mallove, E.F., *Fire from Ice - Searching for the Truth Behind the Cold Fusion Furor*, Infinite Energy Press, USA, 1999.

Manning, J., *The Coming Energy Revolution - The Search for Free Energy*, Avery Publishing Group, Garden City Park, New York, 1996.

Mizuno, T., *Nuclear Transmutation: The Reality of Cold Fusion*, Infinite Energy Press, Concord, NH, 1997.

Von Ward, P., *Our Solarian Legacy - Multidimensional Humans In A Self-Learning Universe*, Hampton Roads Publishing Company, Inc., 2001.

Architecture, Art & Music

Art-Sculpture-Architecture, Harry N. Abrams, Inc., New York, 1989.

Barrett, M., *Aggies, Immies, Shooters, and Swirls - The Magical World of Marbles*, A Bulfinch Press Book, Little, Brown and Company, 1994.

Beardsley, J., *Earthworks and Beyond - Contemporary Art in the Landscape*, Abbreville Press Publishers, New York, 1984.

Davis, D., *Art and the Future - A History/Prophecy of the Collaboration Between Science, Technology and Art*, Praeger Publishers, New York, 1975.

Edwards, E.B., *Pattern and Design With Dynamic Symmetry*, Dover Publications, Inc., New York.

Geck, F.J., *French Interiors & Furniture - The Period of Louis XIII*, Stureck Educational Services, Boulder, CO, 1989.

Geerlings, G.K., *Wrought Iron in Architecture*, Dover Publications, Inc., Mineola, N.Y., 1983.

Ghyka, M., *The Geometry of Art and Life*, Dover Publications, Inc, New York, 1977.

Grieder, T., *Artist and Audience*, Holt, Rinehart and Winston, Inc., Fort Worth, TX, 1990.

Jeans, J., *Science & Music*, Dover Publications, Inc., New York, 1968.

Kaufmann, E., *9 Commentaries on Frank Lloyd Wright*, The MIT Press, Cambridge, MA, 1989.

Kellog, R. and O'Dell. S., *The Psychology of Children Art*, CRM-Random House, 1990.

Leland, N., *Creative Artist - A Fine Artist's Guide to Expanding Your Creativity and Achieving Your Artistic Potential*, North Light Books, Cincinnati, OH, 1990.

McCabe, C.J., *The Golden Door - Artist-Immigrants of America, 1876-1976*, Smithsonian Institution Press, Washington, D.C., 1976.

Moravcsik, M.J., *Musical Sound - An Introduction to the Physics of*

Music, The Solomon Press, Jamaica, NY, 1987.

Norton, F.H., *Ceramics for the Artist Potter*, Central Company, Inc., Cambridge 42, MA, 1956.

Pickover, C.A., *Computers, Pattern, Chaos and Beauty - Graphics From an Unseen World*, St. Martin's Press, New York, 1990.

Pickover, C.A., *Computer and the Imagination - Visual Adventures Beyond the Edge*, St. Martin's Press, New York, 1991.

Pickover, C.A., *Mathematics and Beauty II; Spirals and "Strange" Spirals in Civilization, Nature, Science, and Art*, IBM Thomas J. Watson Research Center, Yorktown Heights, NY, 1987.

Ponter, A. and Pointer, L., *Spirits in Stone - New Face of African Art*, Ukama Press, Glen Ellen, CA, 1997.

Purce, J., *The Mystic Spiral*, Thames and Hudson, New York, New York, 1974.

Rowell, M., *The Planar Dimension - Europe, 1912-1932*, National Endowment for the Arts, Washington, D.C., 1979.

Schaum, J.W., *Chord Speller*, Schaum Publications, Inc., Mequon, WI, 1967.

Schaum, J.W., *Interval Speller*, Schaum Publications, Inc., Mequon, WI, 1966.

Surmani, A., et al, *Essentials of Music Theory*, Alfred Publishing Co., Inc., Van Nuys, CA, 1998.

Treasures of Early Irish Art: 1500 B.C. to 1500 A.D., From the Collections of the National Museum of Ireland, Royal Irish Academy, Trinity College, Dublin, 1977.

Vlach, J.M., *The Afro-American Tradition in Decorative Arts*, Published by The Cleveland Museum of Art, 1990.

Williams, G., *African Designs from Traditional Sources*, Dover Publications, Inc., New York, 1990.

INDEX